Mathematics and Science Education: Assessment, Performance and Estimates

MATHEMATICS AND SCIENCE EDUCATION: ASSESSMENT, PERFORMANCE AND ESTIMATES

CHAD P. ALLERTON
EDITOR

Nova Science Publishers, Inc.
New York

For permission to use material from this book please contact us:
Telephone 631-231-7269; Fax 631-231-8175
Web Site: http://www.novapublishers.com

LIBRARY OF CONGRESS CATALOGING-IN-PUBLICATION DATA

Mathematics and science education : assessment, performance, and estimates / editor, Chad P. Allerton.

p. cm.

Includes index.

ISBN 978-1-60692-313-9 (hardcover)

1. Mathematics--Study and teaching--United States--Evaluation. 2. Science--Study and teaching--United States--Evaluation. I. Allerton, Chad P. II. United States. Dept. of Education. Office of Planning, Evaluation, and Policy Development.

QA13.M3465 2009

510.71'073--dc22 2008052169

Published by Nova Science Publishers, Inc. ✢ New York

CONTENTS

PREFACE

Mathematics and Science Education presents important studies dealing with a segment of education that is closely related to both national security and technological progress in the 21st century.

Chapter 1 - For many years, policymakers have been concerned by the relatively low levels of academic achievement by economically disadvantaged K-12 students in math and science, by the underrepresentation of disadvantaged college students in math and science majors, and by the underrepresentation of people from disadvantaged groups in math and science careers. While racial gaps in math and science test scores narrowed somewhat in the 1970s and 1980s, substantial gaps persisted through the 1990s to the present.

To help address these disparities, the U.S. Department of Education (ED) established a math and science initiative in 1990 within Upward Bound, a federal grant program designed to provide disadvantaged high school students with skills and experiences that will prepare them for college success. The initiative, referred to as Upward Bound Math-Science (UBMS), awards grants to institutions—largely colleges and universities—to operate UBMS projects. These projects were designed to differ from "regular" Upward Bound projects in several respects. To ensure that participants receive an intensive math and science precollege experience, UBMS projects provide instruction that includes hands-on experience in laboratories, computer facilities, and at field sites. Opportunities are also provided to learn from mathematicians and scientists employed at the host institution or engaged in research or applied science in other institutions in the community. A six-week summer program providing intensive instruction in laboratory science and mathematics through precalculus is also offered.

Initially, ED funded 30 UBMS projects. By FY 2004, there were 127 UBMS projects serving 6,845 students at a total cost of $32.8 million. Therefore, the annual cost per student—approximately $4,800—is comparable in cost to regular Upward Bound but much more expensive than other federally-funded precollege programs. More than 80 percent of UBMS projects are hosted by four-year colleges and universities; most of the rest are hosted by two-year colleges (Curtin and Cahalan 2004).

Participants in UBMS must meet the same eligibility requirements as regular Upward Bound students: students must (1) belong to families classified as low-income (taxable income of no greater than 150 percent of the poverty line), or (2) be a potential first-generation college student (neither parent has a bachelor's degree). Some students who participate in UBMS summer programs are referred from regular Upward Bound programs and then return to those programs during the academic year. However, as would be expected, UBMS projects are more likely to consider students' interests in math and science when reviewing applications than are most regular Upward Bound projects (Moore 1997b). While 25 percent of participants are white, most program participants are from underrepresented minority groups: about 60 percent of participants are African American or Hispanic (Curtin and Cahalan 2004).

Chapter 2 - The Program for International Student Assessment (PISA) is a system of international assessments that measures 15-year-olds' performance in reading literacy, mathematics literacy, and science literacy every three years. PISA, first implemented in 2000, is sponsored by the Organization for Economic Cooperation and Development (OECD), an intergovernmental organization of 30 member countries. In 2006, 57 jurisdictions participated in PISA, including 30 OECD jurisdictions and 27 non-OECD jurisdictions.

Each PISA data collection effort assesses one of the three subject areas in depth. In this third cycle, PISA 2006, science literacy was the subject area assessed in depth. The PISA assessment measures student performance on a combined science literacy scale and on three science literacy subscales: *identifying scientific issues, explaining phenomena scientifically,* and *using scientific evidence*. Combined science literacy scores are reported on a scale from 0 to 1,000 with a mean set at 500 and a standard deviation of 100.

This chapter focuses on the performance of U.S. students in the major subject area of science literacy as assessed in PISA 2006.[1] Achievement in the minor subject area of mathematics literacy in 2006 is also presented.[2]

Differences in achievement by selected student characteristics are covered in the final section.

Key findings from the report include:

- Fifteen-year-old students in the United States had an average score of 489 on the combined science literacy scale, lower than the OECD average score of 500. U.S. students scored lower on science literacy than their peers in 16 of the other 29 OECD jurisdictions and sic of the 27 non- OECD jurisdictions. Twenty-two jurisdictions (five OECD jurisdictions and 17 non-OECD jurisdictions) reported lower scores compared to the United States in science literacy.
- When comparing the performance of the highest achieving students—those at the 90th percentile—there was no measurable difference between the average score of U.S. students (628) compared to the OECD average (622) on the combined science literacy scale. Twelve jurisdictions (nine OECD jurisdictions and three non-OECD jurisdictions) had students at the 90th percentile with higher scores than the United States on the combined science literacy scale.

[1] A total of 166 schools and 5,611 students participated in the assessment. The overall weighted school response rate was 69 percent before the use of replacement schools. The fi nal weighted student response rate was 91 percent.

[2] PISA 2006 reading literacy results are not reported for the United States because of an error in printing the test booklets. In several areas of the reading literacy assessment, students were incorrectly instructed to refer to the passage on the "opposite page" when, in fact, the necessary passage appeared on the previous page. Because of the small number of items used in assessing reading literacy, it was not possible to recalibrate the score to exclude the affected items. Furthermore, as a result of the printing error, the mean performance in mathematics and science may be misestimated by approximately 1 score point. The impact is below one standard error. For details see appendix B.

- U.S. students also had lower scores than the OECD average score for two of the three content area subscales (*explaining phenomena scientifically* (486 versus 500) and *using scientific evidence* (489 versus 499)). There was no measurable difference in the performance of U.S. students compared with the OECD average on the *identifying scientific issues* subscale (492 versus 499).
- Along with scale scores, PISA 2006 uses six proficiency levels to describe student performance in science literacy, with level 6 being the highest level of proficiency. The United States had greater percentages of students below level 1 (8 percent) and at level 1 (17 percent) than the OECD average percentages on the combined science literacy scale (five percent below level 1 and 14 percent at level 1).
- In 2006, the average U.S. score in mathematics literacy was 474, lower than the OECD average score of 498. Thirty-one jurisdictions (23 OECD jurisdictions and 8 non-OECD jurisdictions) scored higher, on average, than the United States in mathematics literacy in 2006. In contrast, 20 jurisdictions (4 OECD jurisdictions and 16 non-OECD jurisdictions) scored lower than the United States in mathematics literacy in 2006.
- When comparing the performance of the highest achieving students—those at the 90th percentile—U.S. students scored lower (593) than the OECD average (615) on the mathematics literacy scale. Twenty-nine jurisdictions (23 OECD jurisdictions and six non-OECD jurisdictions) had students at the 90th percentile with higher scores than the United States on the mathematics literacy scale.
- There was no measurable difference on the combined science literacy scale between 15-year-old male (489) and female (489) students in the United States. In contrast, the OECD average was higher for males (501) than females (499) on the combined science literacy scale.
- On the combined science literacy scale, black (non-Hispanic) students (409) and Hispanic students (439) scored lower, on average, than white (non-Hispanic) students (523), Asian (non-Hispanic) students (499), and students of more than one race (non-Hispanic) (501). Hispanic students, in turn, scored higher than black (non-Hispanic) students, while white (non-Hispanic) students scored higher than Asian (non-Hispanic) students.

Chapter 3 - The Nation's Report Card™ informs the public about the academic achievement of elementary and secondary students in the United States. Report cards communicate the findings of the National Assessment of Educational Progress (NAEP), a continuing and nationally representative measure of achievement in various subjects over time.

For over three decades, NAEP assessments have been conducted periodically in reading, mathematics, science, writing, U.S. history, civics, geography, and other subjects. By collecting and reporting information on student performance at the national, state, and local levels, NAEP is an integral part of our nation's evaluation of the condition and progress of education. Only information related to academic achievement and relevant variables is collected.

The privacy of individual students and their families is protected, and the identities of participating schools are not released.

NAEP is a congressionally authorized project of the National Center for Education Statistics (NCES) within the Institute of Education Sciences of the U.S. Department of Education. The Commissioner of Education Statistics is responsible for carrying out the NAEP project. The National Assessment Governing Board oversees and sets policy for NAEP.

Chapter 4 - America has long been innovation's home. It's in our very DNA, born from a desire to be free that was ahead of its time. When faced with a challenge, we invent the answer: from the first telephone to global satellite communications; from the first computer to the World Wide Web; from the Wright Brothers to Neil Armstrong. To Americans, innovation means much more than the latest gadget. It means creating a more productive, prosperous, mobile and healthy society. Innovation fuels our way of life and improves our quality of life. And its wellspring is education.

President Bush made innovation and education top priorities. He worked with Congress to pass the most far-reaching education reform in decades, the No Child Left Behind Act (NCLB). The law brought high standards and accountability to public schools and sparked a mathematics and reading revival in the early grades. The president also increased funding for innovative and intensive reading programs such as Reading First by more than 200 percent since 2001, benefiting more than a million students.

In: Mathematics and Science Education
Editor: Chad P. Allerton

ISBN: 978-1-60692-313-9

Chapter 1

UPWARD BOUND MATH-SCIENCE: PROGRAM DESCRIPTION AND INTERIM IMPACT ESTIMATES*

Robert Olsen, Neil Seftor, Tim Silva, David Myers, David DesRoches and Julie Young

ABSTRACT

For many years, policymakers have been concerned by the relatively low levels of academic achievement by economically disadvantaged K-12 students in math and science, by the underrepresentation of disadvantaged college students in math and science majors, and by the underrepresentation of people from disadvantaged groups in math and science careers. While racial gaps in math and science test scores narrowed somewhat in the 1970s and 1980s, substantial gaps persisted through the 1990s to the present.

To help address these disparities, the U.S. Department of Education (ED) established a math and science initiative in 1990 within Upward Bound, a federal grant program designed to provide disadvantaged high school students with skills and experiences that will prepare them for college success. The initiative, referred to as Upward Bound Math-Science (UBMS), awards grants to institutions—largely colleges and universities—to operate UBMS projects. These projects were designed to differ from "regular" Upward Bound projects in several respects. To ensure that participants receive an intensive math and science precollege experience, UBMS projects provide instruction that includes hands-on experience in laboratories, computer facilities, and at field sites. Opportunities are also provided to learn from mathematicians and scientists employed at the host institution or engaged in research or applied science in other institutions in the community. A six-week summer program providing intensive instruction in laboratory science and mathematics through precalculus is also offered.

Initially, ED funded 30 UBMS projects. by FY 2004, there were 127 UBMS projects serving 6,845 students at a total cost of $32.8 million. Therefore, the annual cost per

---

* Excerpted from U.S. Department of Education, Office of Planning, Evaluation and Policy Development, Policy and Program Studies Service, *Upward Bound Math-Science: Program Description and Interim Impact Estimates*, Washington, D.C., 2007.

student— approximately $4,800—is comparable in cost to regular Upward Bound but much more expensive than other federally funded precollege programs. More than 80 percent of UBMS projects are hosted by four-year colleges and universities; most of the rest are hosted by two-year colleges (Curtin and Cahalan 2004).

Participants in UBMS must meet the same eligibility requirements as regular Upward Bound students: students must (1) belong to families classified as low-income (taxable income of no greater than 150 percent of the poverty line), or (2) be a potential first-generation college student (neither parent has a bachelor's degree). Some students who participate in UBMS summer programs are referred from regular Upward Bound programs and then return to those programs during the academic year. However, as would be expected, UBMS projects are more likely to consider students' interests in math and science when reviewing applications than are most regular Upward Bound projects (Moore 1997b). While 25 percent of participants are white, most program participants are from underrepresented minority groups: about 60 percent of participants are African American or Hispanic (Curtin and Cahalan 2004).

Evaluation of Upward Bound Math-Science

Since 1991, Mathematica Policy Research, Inc., (MPR) has been conducting the National Evaluation of Upward Bound for ED. The centerpiece of this evaluation has been a random assignment evaluation of regular Upward Bound. In 1997, ED added a new component to the evaluation that is focused on UBMS. In 1998, MPR selected a random sample of the students who participated in UBMS between 1993 and 1995 at projects that were still operating at that time. This report constitutes the first of two evaluation reports on UBMS, and it is based on participant surveys and student transcripts collected for this sample between 1998 and 1999 and again between 2001 and 2002. The second report is scheduled for completion in 2006 and will be based on data collected between 2003 and 2005.

The evaluation of UBMS has two components: a descriptive analysis and an impact analysis. The descriptive analysis relies primarily on a survey of project directors to describe the resources available to UBMS projects; the types of institutions that host them; the credentials and demographic characteristics of project staff; recruitment, eligibility, and enrollment of students; student characteristics; and program offerings. The impact analysis is designed to measure the effects of UBMS on (1) performance in high school, especially in math and science courses; (2) postsecondary attendance, persistence and completion; and (3) the likelihood of completing a postsecondary degree in mathematics or a scientific field.

The impact analysis is based on a comparison of UBMS participants with a sample of students that (1) applied to enroll in regular Upward Bound programs in the early 1990s, (2) never participated in UBMS and (3) have been tracked by MPR as part of the national evaluation. This comparison group was selected to ensure that it had similar characteristics to the sample of UBMS participants, and we controlled statistically for the small remaining differences in these characteristics between UBMS participants and the comparison group.

If UBMS participants are more interested or skilled in math and science than the students in the comparison group, the estimated effects of the program may be subject to "selection bias" and may overstate the true effects of participating in UBMS. However, the comparison group we selected was probably the best available short of a randomized control group because the students in the comparison group exhibited the motivation to pursue Upward

Bound services, and our analysis shows that the participant and comparison groups are similar in other ways as well. In addition, we implemented a data collection and analysis plan designed to minimize selection bias (see Chapter III for more details). While a control group from a randomized experiment would have prevented selection bias, the comparison group we selected greatly reduced the cost of the evaluation because we were already collecting data for this group as part of the national evaluation.

Note that the descriptive findings and impact estimates presented in this report describe the operations and effects of the Upward Bound Math-Science Program as it operated in the mid- 1990s. At that time, it was a relatively new program, and some changes have occurred in how UBMS projects operate. In Chapter II, we mention some of these changes as they are reflected in information provided to us by UBMS project directors in a survey of grantees. It is certainly possible that some of the changes in the program since the mid-1990s have influenced the effectiveness of UBMS projects, and the evaluation does not attempt to measure any changes in effectiveness since that time. In this report, we measure the effects of the program on people who participated between 1993 and 1995 and describe the operations of the program at that time.

Report Findings

From our descriptive analysis, we found that UBMS projects:

- ***Provide a large quantity of academic instruction.*** in the summer, the average UBMS project provided a total of 240 hours of academic instruction, and participation in the program is roughly full-time for a six-week period.
- ***Are most active during the summers.*** UBMS projects typically provide services, such as tutoring or study sessions, during the school year, but they provide most of their services during summer residential programs at the colleges or universities hosting the program.
- ***Provide academic enrichment in math and science subjects.*** Many UBMS projects offer courses in algebra II, geometry, precalculus, biology, chemistry, physics and computer software; in contrast, few offer courses in Social Studies (though many offer English courses in addition to their math and science offerings). At most projects, the course work is designed to provide academic enrichment instead of academic remediation.
- ***Provide instruction through a combination of single-subject courses and interdisciplinary instruction.*** While other instructional techniques were used, three out of four projects provided instruction primarily through single-subject academic courses or the combination of these courses with interdisciplinary instruction.

Given the academic services provided by UBMS, it is natural to ask whether participating in UBMS affects the educational outcomes of the students that participate. From our impact analysis, we found that UBMS:

- ***Improved high school grades in math and science and overall.*** UBMS had a positive effect on high school grades, increasing the average GPA in math courses from 2.7 to 2.8, the average GPA in science courses from 2.7 to 2.9 and the average GPA overall.
- ***Increased the likelihood of taking chemistry and physics in high school.*** UBMS increased the likelihood that participants took upper-level science courses in high school, raising the percentage of students taking chemistry from 78 percent to 88 percent and raising the percentage of students taking physics from 43 percent to 58 percent. in contrast, UBMS did not affect coursetaking in advanced math subjects (see Chapter III, Exhibit III.3).
- ***Increased the likelihood of enrolling in more selective four-year institutions.*** UBMS increased the percentage of students that attended four-year colleges and universities from 71 percent to 82 percent. The increase in four-year attendance is particularly pronounced for more selective schools (those rated as "most selective", "highly selective" or "very selective" by the Barron's Guide): UBMS increased the percentage of students that attended more selective four-year colleges from 23 percent to 33 percent (see Chapter III, Exhibit III.4).
- ***Increased the likelihood of majoring in math and science.*** UBMS affected students' choice of major, increasing the percentage majoring (or planning to major) in math or science from 23 percent to 33 percent and decreasing the percentage majoring in a field outside of math or science and the social sciences from 51 to 42 percent. UBMS also seems to increase the percentage of participants majoring in the social sciences (see Chapter III, Exhibit III.6).
- ***Increased the likelihood of completing a four-year degree in math and science.*** UBMS increased the percentage of students that earned a bachelor's degree in a math and science field from 6 percent to 12 percent and decreased the percentage that earned a bachelor's degree outside of math, science, and the social sciences from 20 to 14 percent (see Chapter III, Exhibit III.6). Because 47 percent of participants in our sample were still in college when we interviewed them in 2002, findings related to degree completion should be treated as preliminary, and a final assessment will be presented in a subsequent report.

In addition, we computed separate impact estimates for subgroups defined by sex, race and ethnicity, and prior participation in regular Upward Bound. For some outcomes, we found differences in subgroup impacts that were statistically significant. For example, the effect of UBMS on four-year college attendance was larger for women than for men. However, the number of significant differences between subgroups was relatively small, and there was no obvious pattern to the findings suggesting that particular groups benefited more from UBMS than other groups. Therefore, it is not clear whether the significant subgroup differences are due to chance or to systematic differences in the effects of UBMS on different groups of participants.

To summarize the report's findings, UBMS provides intensive academic instruction in math and science, and our impact estimates suggest that it improves several student outcomes in high school and college. In addition, and consistent with the objectives of the program, preliminary estimates suggest that UBMS participation increases the odds of majoring in math or science. In the next report, we will reexamine the effects on college completion, examine the effects on labor market outcomes, such as employment in the sciences, and weigh the benefits of the program against the costs.

It is tempting to compare the estimated impacts of UBMS to the estimated impacts of regular Upward Bound presented in earlier reports. However, it is important to recognize that the two studies used different methods: while the evaluation of regular Upward Bound is based on an experimental design, the "gold standard" in evaluation research, the evaluation of UBMS is based on nonexperimental methods that may suffer from selection bias, as described earlier. If the estimated effects of UBMS are inflated due to selection bias, then the impression based on our findings that UBMS is more effective than regular Upward Bound might be attributable to differences in the methods used to estimate the impacts instead of differences in the effectiveness of the two programs.

I. Introduction

For many years, policymakers have been concerned by the relatively low levels of academic achievement by economically disadvantaged K-12 students in math and science, by the underrepresentation of disadvantaged college students in math and science majors, and by the underrepresentation of people from disadvantaged groups in math and science careers. National statistics show that while the gaps between minorities'[3] and whites' math and science test scores narrowed somewhat in the 1970s and 1980s, gaps in test scores and other educational outcomes persisted through the 1990s to the present.

- ***Disadvantaged students take fewer math and science courses in high school.*** in the 1991–92 school year, 57 percent of seniors in the lowest socioeconomic status (SES) quartile took a math course, compared with 75 percent of seniors from the highest SES quartile; 37 percent of seniors from the lowest SES quartile took a science course, compared with 61 percent of seniors from the highest SES quartile (U.S. Department of Education 1996b). in 1994, only 58 percent of black high school graduates had completed geometry while in high school, compared with 73 percent of white high school graduates. in the same year, only 13 percent of black and Hispanic graduates had completed the common triad of science courses— biology, chemistry, and physics—compared with 23 percent of white graduates (U.S. Department of Education 1 996a).
- ***Minority college students are less likely to take math and science courses or earn a degree in math or science.*** Ten percent of black college students and 14 percent of Hispanics received credit for calculus or advanced math courses in the late 1 980s, compared with 22 percent of whites. Sixteen percent of blacks and 21 percent of Hispanic college students earned course credits in chemistry, compared with 27 percent of whites, and 8 percent of blacks and 11 percent of Hispanics earned college credit for physics, compared with 18 percent of white

[3] Ideally, socioeconomic measures such as income would be used to define groups, rather than race or ethnicity. For most education outcomes of interest, however, data are not presented on different income groups. Because racial and ethnic minorities are disproportionately lower-income (U.S. Census Bureau 2001:40), data based on race and ethnicity offer a reasonable, albeit imperfect, estimate of economically disadvantaged students' educational experiences.

students (U.S. Department of Education 1994). Because minorities earned fewer college credits in math and science than whites, it is not surprising that they were less likely to earn degrees in those subjects. Black students earned 7 percent of all bachelor's degrees in 1995-96, but just 7 percent of all bachelor's degrees in math and science fields. in the same year, Hispanic students earned 5 percent of all bachelor's degrees, but just 4 percent of all bachelor's degrees in math and science (U.S. Department of Education 1999).[4]

- ***Minorities are less likely than whites to enter careers in math and science.*** Among people who were working in a scientific field in 1995 and had obtained their college degree in the previous five years, only 6 percent were black (National Science Foundation 1995). However, in 1990, around the time those individuals would have been in college, blacks accounted for 14 percent of the U.S. population aged 18-24 years old (Census Bureau 1990a, Census Bureau 1990b).

A. Upward Bound Math-Science Program

To help address these disparities, the U.S. Department of Education (ED) in 1990 established the Upward Bound Math-Science Program (UBMS) within Upward Bound, a federal grant program designed to provide disadvantaged high school students with skills and experiences that will prepare them for college success. UBMS was designed to differ from "regular" Upward Bound in a few key respects. To ensure that participants receive an intensive math and science precollege experience, ED requires UBMS projects to provide instruction that includes hands-on experience in laboratories, computer facilities, and at field sites. Also provided are the following: opportunities to learn from mathematicians and scientists employed at the host institution or engaged in research or applied science in other institutions in the community;[35] involvement with tutors and counselors who are graduate and undergraduate math and science majors; and a six-week summer program consisting of daily course work and activities, instruction in laboratory science and mathematics through precalculus (in addition to foreign language, composition and literature, which are also required offerings at regular Upward Bound projects).

Initially, ED funded 30 UBMS projects. By FY 2004, there were 127 UBMS projects serving 6,845 students at a total cost of $32.8 million. Therefore, the annual cost per student— approximately $4,800—is comparable in cost to regular Upward Bound but much more expensive than other federally funded precollege programs. More than 80 percent of UBMS projects are hosted by four-year colleges and universities; most of the rest are hosted by two-year colleges (Curtin and Cahalan 2004).

UBMS participants must meet the same eligibility requirements as regular Upward Bound participants: they must (1) come from families that are classified as low-income (taxable income not over 150 percent of the poverty line), or (2) be a potential first-generation college student (neither parent has a bachelor's degree). Some students who participate in

---

[4] The following subjects were classified as math or science: biological sciences and life sciences, computer and information sciences, engineering, engineering-related technologies, mathematics, and physical sciences and science technologies.

[5] This requirement may have stemmed from concern that too much math and science instruction in high school is provided by teachers teaching out of their own field.

UBMS are referred from regular Upward Bound programs and then return to those programs during the academic year. However, as would be expected, UBMS projects are more likely to consider students' interests in math and science when reviewing applications than are most regular Upward Bound projects (Moore 1 997b). While 25 percent of participants are white, most program participants are from underrepresented minority groups: about 60 percent of participants are African American or Hispanic (Curtin and Cahalan 2004).

Despite coming from low-income families, the evidence suggests that on average, UBMS serves students who do well in high school and attend college at higher rates than the average low-income student. Data reported by Upward Bound projects suggest that prior to participating in Upward Bound, UBMS participants earned higher grades on average than regular Upward Bound participants (Curtin and Cahalan 2004). In addition, the national evaluation has shown that regular Upward Bound participants would have attended college at much higher rates than the average low-income student even if they had not participated in Upward Bound (Myers et al. 2004).[6] Therefore, the evidence strongly suggests that UBMS serves high school students who are much more likely to attend college than the average low-income student.

B. Evaluation of the Upward Bound Math-Science Program

The legislation establishing Upward Bound authorizes ED to sponsor studies of it, including examinations of program effectiveness. In 1991, ED awarded a contract to Mathematica to conduct the National Evaluation of Upward Bound. This evaluation has several components, but its signature feature is an experiment to measure the effects of participating in regular Upward Bound. We selected a random sample of Upward Bound projects (excluding UBMS projects); for each of these projects, we randomly assigned eligible applicants to a treatment group, which was offered the chance to participate in the program, or a control group, which was not. The evaluation is ongoing, and it was one of the first to use experimental methods to measure the effects of a federally funded education program.

This report presents the results of an evaluation of the Upward Bound Math-Science Program. In 1997, Mathematica completed two reports on UBMS. One provided a descriptive analysis of the program based primarily on site visits to a representative sample of 14 UBMS projects (Moore 1997a). The other provided an assessment of the feasibility of conducting a rigorous evaluation of the effects of UBMS on student outcomes (Myers 1997). When ED awarded a contract to Mathematica in 1997 to extend its evaluation of the effects of regular Upward Bound, it also specified an evaluation of the effects of UBMS. This evaluation consists of two components: a descriptive analysis and an impact analysis. The descriptive analysis relies primarily on a survey of UBMS project directors conducted in the spring of 1998. The analysis is designed to describe the resources available to UBMS projects; the

---

6 While the last report from the national evaluation shows that about 70 percent of regular Upward Bound participants would have attended a postsecondary institution even if they had not participated in Upward Bound (Myers et al. 2004), only 53 percent of 1992 high school graduates from the lowest SES quartile attended a postsecondary institution by 1994 (U.S. Department of Education 1997).

types of institutions that host them; the credentials and demographic characteristics of project staff; recruitment, eligibility, and enrollment of students; student characteristics; and a description of the program, including its goals, academic orientation, instructional methods and the intensity and quantity of the services provided.

The UBMS impact study is designed to measure the effects of participating in UBMS on college enrollment, choice of major, and other outcomes for students who participated during the summer of 1993, 1994, or 1995.[7] Conceptually, the study contrasts how participants fared with how they would have fared if they had not participated in UBMS. We compared UBMS participants with eligible applicants to the regular Upward Bound projects participating in the national evaluation. From this pool, we systematically selected a matched comparison group of students who were as similar as possible to UBMS participants in terms of characteristics and experiences that could potentially predict later outcomes. These characteristics included demographics—such as sex, race, and ethnicity—and prior academic achievement such as grade point average and math and science courses taken in 9th grade. The key difference was that the matched comparison students did not participate in UBMS.

The selection of matched comparison students also took into account experiences in other precollege programs, in particular regular Upward Bound. Because regular Upward Bound is an intensive program that can influence high school achievement and postsecondary outcomes (Myers and Schirm 1999; Myers et al. 2003), it is important to account for exposure to regular Upward Bound when estimating how UBMS participants would have fared if they had not participated in UBMS. For UBMS participants who had previously participated in a regular Upward Bound program—perhaps during the academic year—we selected comparison students who had also participated in the regular Upward Bound.[8] For UBMS participants who had not participated in regular Upward Bound, we selected comparison students who did not participate in regular Upward Bound.[9]

Several data sources play a key role in the impact analysis. Baseline characteristics were collected for comparison group members through the baseline survey for the evaluation of regular Upward Bound; baseline information on many of the same characteristics was collected for UBMS participants through a follow-up survey conducted in 1999.[10] This follow-up survey was used to collect information about educational outcomes for UBMS participants, and a similar survey was used to collect analogous information for comparison students. Finally, secondary and postsecondary transcripts were collected for both types of students to assess academic achievement.

---

[7] Because the sample was not selected until 1998, we restricted the sample to participants at UBMS projects that were still operating that year: obtaining lists of participants from programs that were no longer operating in 1998 would have been nearly impossible.

[8] These comparison students were selected from the treatment group for the evaluation of regular Upward Bound. For a more thorough discussion of how the treatment group was selected, see Myers et al. (2004), Appendix A.

[9] These comparison students were selected from the control group for the evaluation of regular Upward Bound. For a more thorough discussion of how the control group was selected, see Myers et al. (2004), Appendix A.

[10] While the 1999 survey was conducted four to six years after our sample had participated in the program, most of the baseline information collected—including sex, race, and ethnicity—is time-invariant.

C. Overview of This Report

The remainder of this report is organized as follows. Chapter II describes the operation of the UBMS program. Chapter III presents findings from the impact analysis.

II. The Operation of the Upward Bound Math-Science Program

To interpret information on the impacts of UBMS, it is necessary to understand what the program entails. This chapter describes key features of the operations of UBMS projects, including the characteristics of host institutions and staff, projects' recruitment practices and enrollment levels, participants' characteristics and projects' goals and services. For context, this chapter presents comparable information on the operations of regular Upward Bound when possible.

The primary data source for this chapter is a survey of UBMS projects conducted in the spring of 1998. The survey sample consisted of all 81 projects operating at the time, and 74 of the 81 projects responded to the survey.[11,12] the survey requested information about program operations in two separate years—(1) 1994 (in the middle of the period over which our sample was participating in UBMS) and (2) 1998 (the year prior to the survey)—but some questions were specific to 1998. When possible, we focus our analysis of program operations on 1994 to facilitate comparisons with regular Upward Bound projects operating in 1993, as reported in Fasciano and Jacobson (1997) and to describe the programs that served the same cohorts of participants for whom we measured the impacts of the program (see Chapter III).[13] To augment the information provided by the survey of UBMS projects, we also use information from case studies and annual performance reports (Moore 1 997b).

The findings in this chapter indicate that UBMS projects provide intensive academic enrichment to disadvantaged high school students in math and science using staff with strong academic credentials in those subjects. Some of the features that make UBMS projects distinctive, even from regular Upward Bound projects, are: (1) high levels of annual funding per student and low student-teacher ratios, (2) recruiting strategies that attract students from wide geographic areas, (3) service provision that is heavily concentrated in residential programs during the summer, (4) course offerings that focus on math and science relative to other subjects, (5) academic preparation over nonacademic college preparatory activities, and (6) academic enrichment over remediation. The remainder of the chapter provides a description of UBMS and an assessment of its distinctive features.

---

[11] We did not adjust (weight) for survey nonresponse, reasoning that the number of nonrespondents was low enough to eliminate any serious concerns about data representativeness. Also, rarely did more than three UBMS projects fail to respond to any particular item on the questionnaire.

[12] We excluded the one project that reported serving only veterans in 1998. Note that veterans' projects were also excluded from the survey of regular Upward Bound grantees, so the comparisons that are made in this chapter between the two types of Upward Bound programs are based on Upward Bound projects that did not exclusively serve veterans.

[13] Unless noted otherwise, the results for 1998 were generally similar to those for 1994.

A. Project Hosts and Staff

The impacts of UBMS projects on student outcomes may depend on the types of institutions that host them and the people they hire to serve as instructors and other staff. In this section, we describe the types of institutions that host UBMS projects and the staff that provide services to program participants.

1. Host institutions: Two- and Four-Year Colleges and Universities

The types of institutions that host a UBMS project may influence where students attend college. Most Upward Bound programs are hosted by either two- or four-year postsecondary institutions. Evidence from the national evaluation of regular Upward Bound suggests that participation at projects hosted by four-year colleges raises the probability of attending a four-year college, and participation at projects hosted by two-year colleges raises the probability of attending a two-year college (Myers et al. 2002b). Therefore, the types of institutions that host UBMS projects may influence the types of postsecondary institutions that program participants subsequently attend.

Nearly nine out of ten UBMS projects operating in the mid–1990s were hosted by four-year colleges, a substantially higher proportion than among regular Upward Bound projects (see Exhibit II.1). Four-year colleges may find it easier than other potential host institutions to meet some of ED's guidelines for UBMS, including offering hands-on experience in laboratories and computer facilities, opportunities to learn from mathematicians and scientists engaged in research or applied science and involvement with tutors and counselors who are graduate and undergraduate students in math and science.

2. Summer Program Staff and Project Director

UBMS projects are directed by highly educated individuals and staffed by people with strong credentials in math and science. These staff have responsibility for a relatively small number of students, which may provide opportunities for individual instruction. At the typical project, the project director and staff can provide same-race role models for many of the students they serve. The sections below provide more detail on our findings concerning staff size and composition by job title, staff credentials, and the racial composition of project staff.

a. Staff Size and Composition by Job Title

In 1994, UBMS projects had an average of 24 staff members, comprising roughly eight instructors, five resident counselors, four mentors, three tutors, two administrators, one academic or guidance counselor and one clerical staff member.[14] Overall, the average student-staff ratio in summer 1 998 was 2: 1, with a range from about 1 : 1 to 5: 1.

These findings, combined with findings from Moore (1997b), suggest that student-staff ratios are typically lower in UBMS projects than in regular Upward Bound projects. The survey of grantees did not collect information on the number of full-time-equivalent (FTE) staff, but the information available suggests that UBMS projects typically maintain student-staff ratios that are substantially lower than in regular Upward Bound. Moore (1997b) found that 14 randomly selected UBMS projects visited in summer 1996 had an average of 2.6 students per FTE staff, including administrators, and 8.2 students per FTE instructional staff

(Moore 1 997b, Exhibit II.6). In contrast, tabulations from the data used by Fasciano and Jacobson (1997) indicate that in summer 1992, regular Upward Bound projects had more students per staff member—5.1 students per FTE staff and 13.6 students per FTE instructional staff (Moore 1997b, Exhibit II.6).

Exhibit II.1. Types of Institutions that Hosted UBMS Projects, 1995

Type of institution	Upward Bound Math-Science	Regular Upward I
Four-year college or university	88%	68%
Two-year college	11	28
Other institution	2	4

Note: Percentages may not sum to 100 percent due to rounding.
Source: Moore 1997b, Exhibit II.1, p. 15.

b. Credentials

At the average UBMS project in 1998, most staff were highly educated and had educational backgrounds in math and science. About one-quarter had attended some college but not obtained a degree, another quarter had obtained a bachelor's or associate's degree (mostly bachelor's degrees) and the rest had done graduate work or obtained a graduate degree. Most staff members without undergraduate degrees were undergraduate students who served as UBMS mentors, tutors and resident counselors while working toward bachelor's degrees in math, science, or education. Approximately two out of five staff members had their highest degree in science or the social sciences (31 percent) or math (10 percent); additionally, most staff members without a degree were working toward a bachelor's degree in math, science, or education.

Most instructors at the average UBMS project had experience teaching math or science. During the school year, most instructors were either high school teachers (41 percent) or postsecondary teachers (31 percent); one-fifth were graduate students (14 percent) or undergraduates (6 percent). Moreover, at the typical project, two-thirds of the high school teachers and three-quarters of the postsecondary teachers taught in a math or science field.

The professional and educational backgrounds of UBMS project directors provide insight on their credentials to direct projects. One-fifth of the directors were faculty members at the host institution or another college, roughly the same percentage as in regular Upward Bound. In 1998, two-thirds of the directors held a master's degree and one-fifth held a doctorate. About half had their highest degree in education, and less than one-fifth had their highest degree in engineering, mathematics, or physical sciences.[15] Although UBMS project directors were less likely than program staff as a whole to have a background in math or science, this may not be surprising since subject area expertise is probably less important for administrators than for other staff. Compared with regular Upward Bound project directors, UBMS project directors were more highly educated. They were, for example, twice as likely to have a doctorate.

[14] By 1998 the average MSC had almost 26 staff, including 9 instructors.
[15] An additional 12 percent had their highest degree in the social sciences.

c. Race and Ethnicity

At about 9 out of 10 UBMS projects operating in 1998, one racial or ethnic group accounted for a majority of the staff (see Exhibit II.2). In many cases this pattern may have reflected a conscious strategy, also used in regular Upward Bound, to provide minority students with same- race role models. For example, at 21 UBMS projects, a majority of the staff members were black; at 18 of these, a majority of the students were also black.

Exhibit II.2. Predominant Racial and Ethnic Group for UBMS Staff, Summer 1998

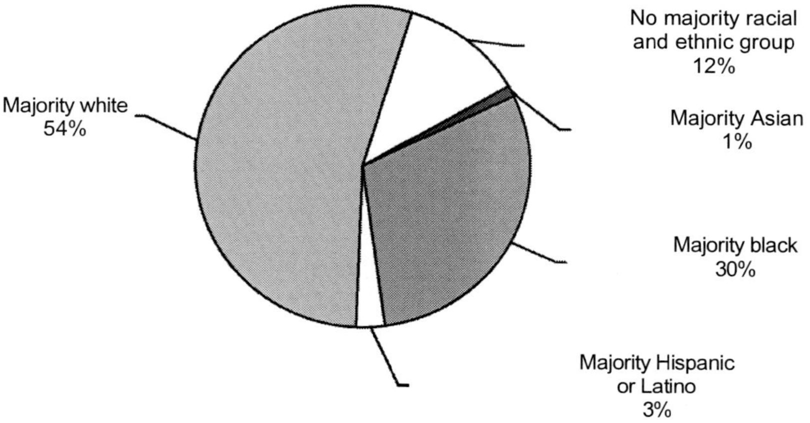

Note: Percentages may not sum to 100 percent due to rounding. The 12 percent of projects with no majority racial and ethnic group included 6 percent of projects with a plurality of blacks and 4 percent with a plurality of whites; the remainder had no plurality.
Source: 1998 survey of UBMS projects.

The racial and ethnic profile of UBMS project directors was similar to that of UBMS project staff (see Exhibit II.3). Both staff and project directors were nearly evenly split between white and nonwhite; project directors were slightly more likely to be black and less likely to be Asian than other staff. The race and ethnicity of the UBMS project director often matched that of the predominant student racial and ethnic group: more than three-fourths of the minority directors headed programs where students from the same group constituted a plurality of participants.

B. Eligibility, Recruitment and Enrollment, and Student Characteristics

To shed light on the types of students that participate in UBMS, we examine the eligibility criteria that students must meet and the recruiting strategies that UBMS projects use to attract students. We also examine the characteristics of participants as reported by project directors.

Exhibit II. 3. Racial and Ethnic Distribution for UBMS Staff and Project Directors, Summer 1998

Race and Ethnicity	Staff	Director
White	49%	49%
Black	33	39
Hispanic or Latino	7	7
Asian	4	1
American indian or Alaskan Native	2	4

Note: Percentages may not sum to 100 percent because (1) people may fall into multiple categories, (2) the Pacific Islander category was excluded from the exhibit because only nine staff members nationwide fell into this category and (3) some staff members may not have been classified by race or ethnicity.
Source: 1998 survey of UBMS projects.

1. Eligibility

UBMS projects have to meet the same federal rules as regular Upward Bound projects concerning the composition of participants. At each project, at least two-thirds of the participants must be both low-income and potential first-generation college students; the remaining students must meet either of these two criteria. At the average project in summer 1998, about 77 percent of students met both of these eligibility criteria, about 14 percent were first-generation only and about 9 percent were low-income only, very similar to the distribution in regular Upward Bound during 1992-93. In addition, UBMS and regular Upward Bound projects are only allowed to serve students who have completed 8th grade.

Most UBMS projects also adopt additional student eligibility criteria for enrollment in the program—for example, requirements about grade level, school course work, or recommendations. In 1994, over three-fourths of projects required students to have finished 9th grade; a few projects required 10th grade completion. In addition, nearly all UBMS projects required a teacher recommendation and completion of at least one high school course in math or science, and applicants enrolled in regular Upward Bound commonly needed a recommendation from the director. Finally, about 30 percent of UBMS projects prohibited students from returning from the previous summer's program, and almost half prohibited students from returning unless they met certain criteria.[16]

2. Recruitment and Enrollment

To find a pool of potentially eligible applicants, UBMS projects focused mainly on other precollege programs or secondary schools (see Exhibit II.4). Among UBMS projects operating in summer 1994, nearly all recruited from regular Upward Bound projects, while substantial majorities also recruited from Talent Search projects and from middle or high schools directly. However, it is important to note that while almost all UBMS projects

[16] These practices changed dramatically by 1998, when only 7 percent of UBMS projects prohibited all students from returning from the prior summer, and 71 percent prohibited students from returning unless they met certain criteria. The questionnaire did not address the specific types of criteria that projects imposed.

recruited from regular Upward Bound projects in 1994, data from the evaluation suggest that fewer than one in five UBMS participants had previously participated in regular Upward Bound.

Exhibit II.4. Targets for Recruiting by UBMS Projects, 1994

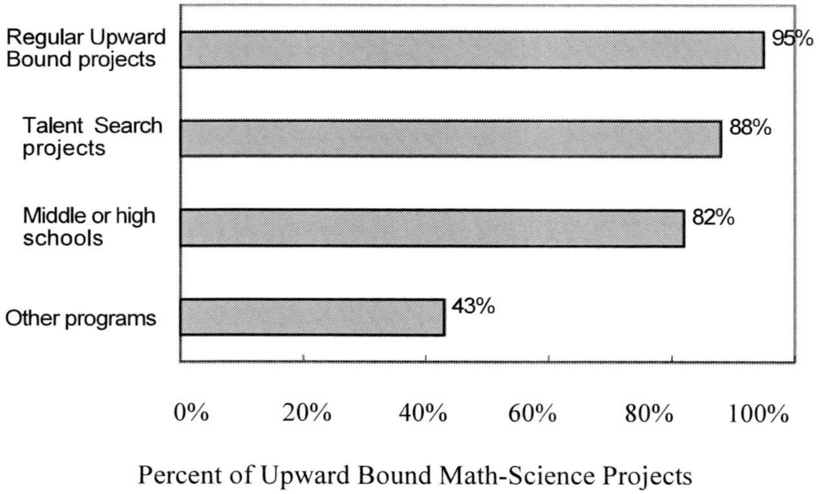

Source: 1998 survey of UBMS projects

Exhibit II.5. Geographic Scope of Recruiting by UBMS Projects, 1994

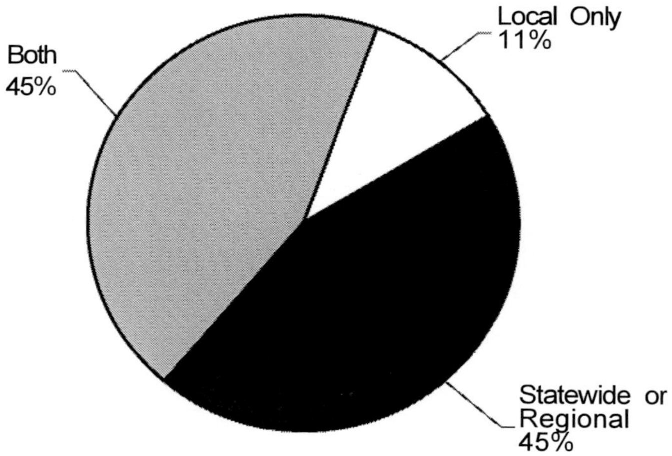

Note: Percentages may not sum to 100 percent due to rounding.
Source: 1998 survey of UBMS projects.

Historically, UBMS projects have cast a wide net in recruiting students beyond the local areas of the host institutions. In 1994, only about one in ten UBMS projects recruited exclusively from a specific and typically local set of feeder schools or Upward Bound projects; the rest recruited from state-wide or regional lists of schools and programs (see Exhibit II.5). By 1998, however, the percentage of UBMS projects that recruited exclusively from a specific set of feeder schools or Upward Bound projects had tripled. Therefore, it appears that over time, more UBMS projects are taking local recruitment strategies like regular Upward Bound projects.

By design, UBMS projects are smaller than regular Upward Bound projects. Through recruitment, UBMS projects received an average of 108 applications for the summer of 1994, ranging from a low of 50 to a high of 300, and they enrolled between 40 and 53 students. In contrast, regular Upward Bound programs enrolled an average of about 75 students in the mid- 1990s (Moore 1997b).

3. Student Characteristics

In our 1998 survey of UBMS projects, project directors provided information on the distribution of students participating in their projects by sex, race, grade level, and place of residence:[17]

- ***Sex.*** At the average UBMS project, like the average regular Upward Bound project, 60 percent of students were female.[18] However, this varied considerably across UBMS projects from a low of 25 percent to a high of 78 percent.
- ***Race.*** On average, UBMS projects served an ethnically diverse group of students: 42 percent black, 27 percent white, 15 percent Hispanic, 8 percent Asian, 5 percent American indian and 1 percent Native Hawaiian or other Pacific Islander. However, most UBMS projects serve participants where one racial and ethnic group constituted a majority (see Exhibit II.6).[19] Furthermore, some UBMS projects exclusively served students from a single racial or ethnic group. For example, six UBMS projects reported that all of its participating students were black.
- ***Grade level. T****he eligibility guidelines discussed above, along with other factors, can affect the distribution of students across different grade levels. At the average project, 29 percent of participants were entering 12th grade, 37 percent were entering 11th grade, 27 percent were entering 10th grade and 6 percent were entering 9th grade. These exhibits suggest that on average, UBMS projects serve students who are slightly closer to graduation than is the case at regular Upward Bound projects.[20] However, there was substantial variation in the grade level distribution of participants across projects in 1998. for example, one UBMS project reported that all of its participants were entering 12th grade, while four reported that none were rising seniors.[21]

---

[17] Most of these questions were focused on 1998 participants, so the analysis in this section is focused on 1998 instead of 1994.

[18] Data based on all UBMS projects operating in 1998 and all regular Upward Bound projects operating in 1992.

[19] In comparison, at 87 percent of regular Upward Bound projects operating in 1992–93, one racial and ethnic group accounted for a majority of participants.

[20] In regular Upward Bound during the summer of 1992, 20 percent of participants were entering 12th grade, 32 percent were entering 11th grade, 31 percent were entering 10th grade, and 16 percent were entering 9th grade.

[21] In addition, the proportion of rising juniors ranged from 0 to 72 percent, and the proportion of rising sophomores ranged from 0 to 60 percent.

- ***Place of residence.*** Given that most projects recruited across the state or region, it is not surprising that many UBMS participants came from outside the grantee's local city or town. At the average project, only about 25 percent of the students were locals. As we would expect, projects that recruited only from a set of local schools or regular Upward Bound projects served considerably higher percentages of students from the local area than other UBMS projects.

Exhibit II.6. Predominant Racial and Ethnic Group for Participants at UBMS Projects, Summer 1998

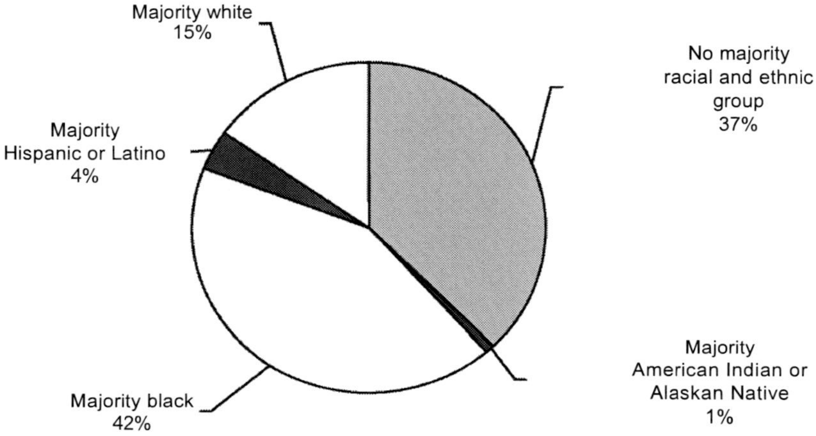

Note: Percentages may not sum to 100 percent due to rounding. The 37 percent of UBMS projects with no majority racial and ethnic group of students included 17 percent with a plurality of white students, 11 percent with a plurality of Hispanics, 6 percent with a plurality of blacks, 3 percent with a plurality of American indians and 1 percent with a plurality of Asians.

Source: 1998 survey of UBMS projects.

Moore (1997b) described UBMS participants as a more select group than regular Upward Bound participants based on having earned somewhat higher grades and having greater interest in math and science (pp. 23, 26) prior to participating in Upward Bound. Discussions with UBMS and regular Upward Bound staff revealed that UBMS participants were typically considered "more serious about school" than regular Upward Bound participants (Moore 1 997b, p. 26).

C. Program Description

In a college-like setting, UBMS projects offer academic enrichment in math and science to improve student achievement in those subjects and expose students to math and science careers. In this section, we describe the following features of UBMS projects in more detail: the setting in which these projects provide services; the goals, academic orientation, academic offerings, and instructional approaches of these projects; and the intensity and quantity of services the UBMS projects provide.

1. Setting

As described earlier, UBMS projects are typically hosted by two- and four-year postsecondary institutions (see Section A.1). Most UBMS projects are hosted by four-year colleges and universities, and most of these institutions have dormitories to house their students. These dormitories are often available in the summers to house participants of summer programs hosted by these institutions.

UBMS projects typically exposed participants to a college setting during the summer program by housing them in the college dormitories. Virtually all the UBMS projects we surveyed (100 percent in 1994, 97 percent in 1998) offered a residential component to their summer programs, compared with 87 percent of regular Upward Bound programs in 1992 (Moore 1997a). At almost all UBMS projects, students lived in the dormitories for the entire summer program, which lasted about six weeks on average.[22] Therefore, for six weeks, participants lived on campus like many college students do during the academic year.

2. Goals of the Program

As mentioned in Chapter I, the general objective of the Upward Bound Math-Science program is to prepare participating students for postsecondary programs leading to careers in math and science. Seven out of ten UBMS projects operating in 1994 rated "academic performance in math and science" as their most or second most important goal (see Exhibit II.7). The focus on academic improvement was similar to the focus of regular Upward Bound projects operating around the same time.[23]

However, two goals that regular Upward Bound projects considered moderately important were not considered important by UBMS projects. First, only 13 percent of UBMS projects reported that one of their top two goals was fostering students' personal skills (e.g., goal orientation, ability to adapt to new settings), compared with 31 percent of regular Upward Bound programs. Second, none of the UBMS projects cited improving students' access to financial aid as one of their top two goals, compared with 35 percent of regular programs.

3. Academic Orientation

UBMS projects try to provide academic enrichment beyond what students are exposed to in school (see Exhibit II.8). Very few projects emphasized remedial instruction in 1994.[24] While about one in four regular Upward Bound programs emphasized the provision of remedial instruction, fewer than one in ten UBMS projects reported doing the same.

The focus of UBMS on academic enrichment over remediation is consistent with the types of students served by the program. As described earlier, findings in Moore (1997b)

---

[22] Only 4 percent of UBMS projects in 1994 had a residential component shorter than the summer program, but by 1998, the rate had increased to 11 percent.

[23] Eighty-seven percent of regular Upward Bound projects rated "academic improvement" as their most or second most important goal in 1993. If the regular Upward Bound grantee survey had also listed "academic improvement in math and science," it is possible that some respondents would have cited that as one of their top two goals: Fasciano and Jacobson (1997) characterized 37 percent of regular Upward Bound projects as having a strong emphasis on math and science.

[24] By 1998, the relative focus on enrichment was even greater, with 83 percent of UBMS projects citing enrichment as a major emphasis and only 1 percent citing remediation.

indicate that, on average, UBMS participants probably had less need for remedial support than regular Upward Bound participants.

Exhibit II.7. Most Important Goals of UBMS Projects, Summer 1994

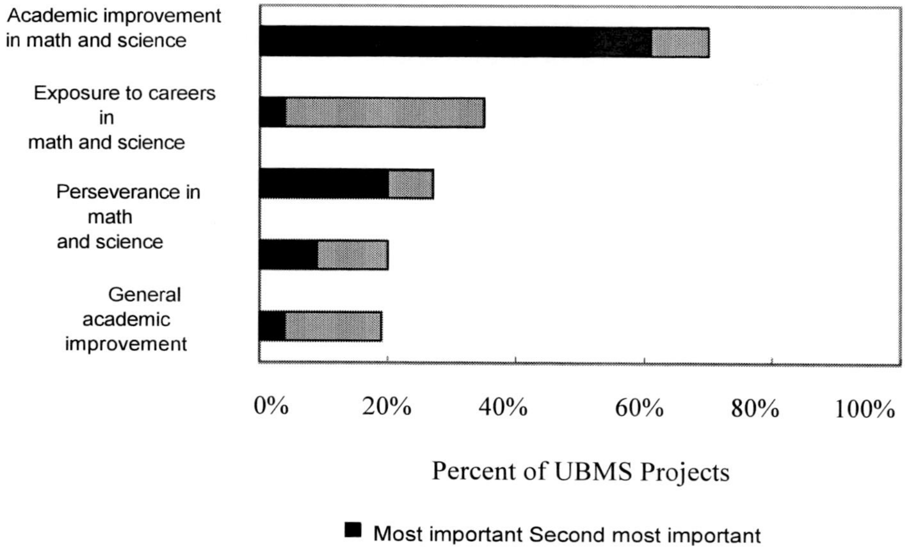

Source: 1998 survey of UBMS projects

4. Summer offerings

In accordance with program guidelines, UBMS projects offered instruction in a diverse array of academic subject areas (see Exhibit II.9). Seventy-five percent or more of these projects in 1994 offered instruction in the following subjects: writing and composition, algebra II, geometry, precalculus, computer applications and software use, biology, chemistry and physics. The average total number of offerings in 1994 was about 14, with a range of 2 to 22. The average number of offerings in math and science was about 7, with a range of 1 to 11.[25] Thus, on average, math and science courses accounted for roughly half of UBMS projects' total offerings.

---

[25] It is not clear why some UBMS projects apparently had such a small number of instructional offerings. One possibility is that their instructional approach centered on interdisciplinary courses or large projects that covered multiple subject areas. However, only 3 of the 12 projects that reported five or fewer math and science offerings in 1994 also reported that their primary instructional method was interdisciplinary courses or large projects and

Exhibit II.8. Academic Orientation of UBMS Projects[a]

Major Emphasis	UBMS Summer 1994	Regular Upward Bound Summer 1 992[b]
Support—instruction that parallels what students are taught in regular school courses	33%	55%
Remediation—instruction that concentrates on fundamental concepts and skills that were taught in earlier grades	6	23
Enrichment—instruction in concepts and material beyond what students are exposed to in regular school classes	73	69

[a] Percentages do not sum to 100 because grantees were allowed to rate more than one approach as a major emphasis.

[b]Excludes summer bridge programs for Upward Bound participants who have just graduated from high school.

Sources: 1998 survey of UBMS projects, Fasciano and Jacobson 1997.

UBMS projects clearly differed from regular Upward Bound projects in their relative emphasis on certain subjects. First, as expected, they concentrated their offerings more on math and science. Although UBMS projects were no more likely than regular Upward Bound projects to offer certain math or science courses (for example, algebra II, geometry precalculus, calculus, biology and chemistry), UBMS projects were much less likely to offer instruction in areas outside of math and science, such as social science or history courses or electives or nonacademic courses. Second, consistent with their greater emphasis on enrichment than on remediation, UBMS projects were less likely than regular Upward Bound programs to offer low- end courses such as reading comprehension and vocabulary, pre-algebra, and earth science (see Exhibit II.9).

To help prepare students for a postsecondary education and post-collegiate careers in math or science, UBMS projects also offered a range of support services and activities (see Exhibit II.10). Among the most common activities were field trips (e.g., to math or science facilities) and assistance with college and financial aid applications. The average number of these non-instructional offerings in 1994 was about 10, with a range of 3 to 15.

UBMS projects were substantially less likely than regular Upward Bound projects to offer services focused on preparing for college. Many regular Upward Bound projects would have provided these services during the academic year: both regular Upward Bound and Math Science programs focus on academics during the summer. Because many UBMS participants participated in other precollege programs during the academic year, UBMS project staff could have reasonably expected that those other programs were assisting students in preparing for college.

experiments. A second possibility is underreporting of instructional offerings by project directors.

Exhibit II.9. Instruction Offered by UBMS Projects, by Subject Area[a]

	UBMS Summer 1994	Regular Upward Bound 1992[b]
English/Language Arts		
Writing/Composition	93%	100%
Literature	60	83
Reading Comprehension and	65	98
English as a Second Language	13	11
foreign Language	54	35
Other	9	13
Mathematics		
Pre-Algebra	36	82
Algebra I	69	96
Algebra II	81	95
Geometry	80	95
Precalculus	80	80
Calculus	52	58
Statistics[c]	17	c
Trigonometry[c]	7	c
Other	9	24
Computers		
Programming	43	47
Applications/Software Use	85	79
internet/Web Page Design[c]	7	c
Other	7	6
Science		
Physics	76	63
Biology	87	89
Chemistry	81	81
Earth Science	48	66
Other	15	19
Social Science/History		
History	11	47
Geography	9	24
Sociology	4	17
Psychology	8	15
Government/Civics	9	40
Other	8	13
Electives/Non-Academic Courses		
Performing Arts	31	53
Art	26	53
Journalism	28	52
Speech/Public Speaking	48	59
Physical Fitness	56	69
Other	6	26

[a] UBMS projects offer instruction in many areas besides math and science, either to meet regulatory requirements or simply to ensure that their program will interest and benefit students in many ways.

[b] 1992 non-bridge summer programs or 1992-93 academic year.

[c] Neither survey listed statistics, trigonometry, or internet or Web page design, but enough project directors specified them under "other courses" that we present data on these courses.

Sources: 1998 survey of UBMS projects, Fasciano and Jacobson (1997:39).

Exhibit II.10. Noninstructional Services Offered by UBMS Projects

	UBMS Summer 1994	Regular Upward Bound 1992[a]
College preparation/skills		
Campus visits	74%	98%
Adjusting to college living	98	92
ACT/SAT preparation	65	97
P SAT/PLAN or PACT preparation	24	73
Help with financial aid or scholarships[b]	80	100
Assistance with college applications	78	99
Assistance with financial aid applications	72	100
Career/employment assistance		
Site visits to employers[c]	65	59
On-campus (employers or career representatives)	63	78
Project-related work experience	49[d]	
JTPA job	0	d
Work-study job	4	d
Math or science internships	24	d
Job through other partnerships	7	d
Field trips to		
Academic science or math facilities	98	e
Non-academic science or math facilities	94	e
Conduct math- or science-related field work	85	e
Other	93	e

[a]1992 nonbridge summer programs or 1992-93 academic year.

[b] In the regular Upward Bound grantee survey, this item was phrased, "Identify sources of financial aid."

[c] In the regular Upward Bound grantee survey, this item was phrased, "Site visit to employers or job shadowing." [d]Although the regular Upward Bound grantee survey had three separate items about JTPA, work-study and other partnerships, the results were reported only in the aggregate, and it did not ask about math or science internships (Fasciano and Jacobson 1997:54).

[e]the regular Upward Bound grantee survey asked about field trips of varying lengths, not the destinations.

Sources: 1998 survey of UBMS projects, Fasciano and Jacobson (1997:54).

5. Academic-Year Offerings

While UBMS projects also provided services to students during the academic year in the mid-1990s, these services were minimal compared to UBMS summer services and also typically far less numerous and less intense than academic year services provided in regular Upward Bound. During the 1994-95 academic year, about one-third of UBMS projects provided tutoring or study sessions, and just over half provided assistance with college applications (see Exhibit II.1 1). The average UBMS project provided about three types of these services during 1994- 95.[26] Not surprisingly, geographic proximity to their participating

[26] By 1998-99, the percentage of UBMS projects regularly providing these services had increased substantially. For example, the percentage providing tutoring or study sessions rose from 81 percent to 93 percent, and the percentage providing assistance with college applications rose

students influenced whether UBMS projects provided certain services during the academic year. UBMS projects were substantially more likely to provide tutoring and workshops during the academic year if a relatively large percentage of their participants lived in the same city or town as the program host.[27]

Exhibit II.11. Academic Year Services Offered by UBMS Projects, 1994-95

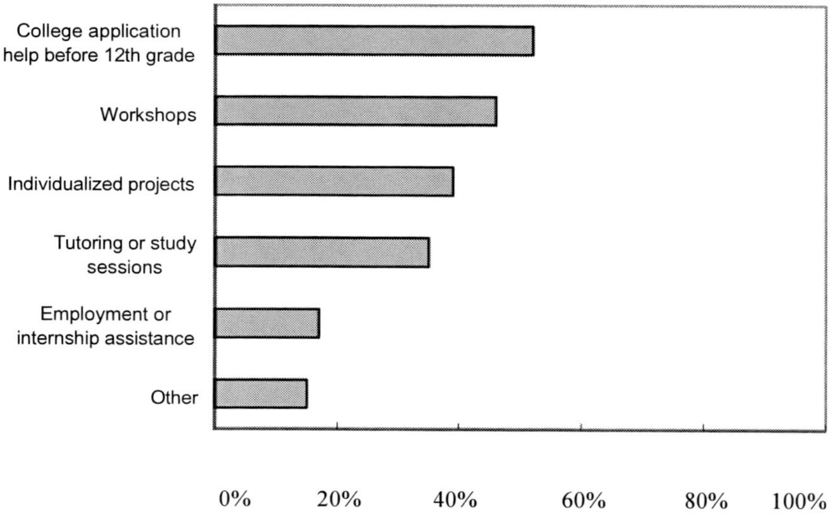

Source: 1998 survey of UBMS projects

6. Instructional Approaches

In 1994, the most common instructional approach taken by UBMS projects was the provision of instruction through courses in separate subjects. Four out of five UBMS projects offered courses in separate subjects (see Exhibit II.12, Panel A). In three out of four UBMS projects, the primary method of instruction was either the provision of these courses (37 percent) or the combination of these courses with interdisciplinary courses (also 37 percent, see Exhibit II.12, Panel B).

However, UBMS projects frequently employed other instructional methods. The majority of projects (63 percent) offered interdisciplinary courses, and in a large minority (35 percent), at least some students worked on a large project or experiment that spanned multiple academic subject areas (see Exhibit II.12, Panel A).

UBMS projects vary considerably in how they sort students into classes or groups. in the summer of 1994, about half of projects placed their students in instructional groups based on proficiency level (37 percent) or grade level (16 percent). About one-fourth placed students with diverse proficiency levels in the same group to facilitate learning (presumably the

---

from 52 percent to 78 percent. This probably reflects the establishment of more locally oriented UBMS projects.

[27] For example, more than three-quarters of UBMS projects with a relatively large percentage of participants from the local area (above the median) offered tutoring during the academic year, versus less than half of UBMS projects with a relatively small percentage (below the median).

learning of less proficient students), and the remaining projects grouped students by their interests or in some other way.

UBMS participants do not spend most of their time in traditional lecture-style classes. At the average project during the summer of 1994, only one-fourth of the time was spent in lecture- style classes like those offered in most schools. The remaining time was spent in small group, teacher-led instruction (32 percent), laboratories (29 percent), computer-based instruction (12 percent), and other settings (4 percent).

D. Intensity and Quantity of Services

The services that UBMS projects offer and the length of their summer residential summer programs suggest that these projects offer intensive programs that provide students with a "large dose" of services, at least for one summer. Furthermore, larger doses of effective services may yield larger impacts than smaller doses, as we found for regular Upward Bound (Myers et al. 2004). Summary measures of program intensity presented in this section indicate that UBMS projects offer an intensive program that might be expected to improve the math and science preparation of program participants.

UBMS is a resource-intensive program. Program grants to UBMS projects provided an average of approximately $4,800 per student in FY 2004 (see Exhibit II.13). This is comparable to funding for regular Upward Bound—approximately $4,500 per student—and much more expensive than most precollege programs. UBMS funding supports an extensive package of instruction and services, as described earlier in the chapter.

Participants devote a substantial amount of time to the program, and most of this time is spent on academics. At the average UBMS project in the summer of 1994, students spent about 29 hours per week receiving instruction and almost 11 hours per week on tutoring and homework. Thus, participating in UBMS over the summer is somewhat like having a full-time job requiring 40 hours per week. Because the vast majority of UBMS summer programs last six weeks,[28] participants at the average project spend 240 total hours on academics during the summer.

The amount of time devoted to core program activities varied across UBMS projects. For example, four projects reported that students spent over 40 hours per week in instruction alone, while a handful of projects reported estimates of less than 10 hours per week. Estimates of time spent on tutoring or homework ranged from 0 to 20 hours per week. And the estimates of average total time spent on academics during the summer varied considerably, from under 100 hours at some projects to over 340 hours at some others.

[28] Only five UBMS projects reported a program of a different length—three had a five-week session, one had a seven-week session, and one had an eight-week session.

Exhibit II.12. Instructional Methods Used by UBMS Projects, 1994

A. Use of Different Instructional Methods

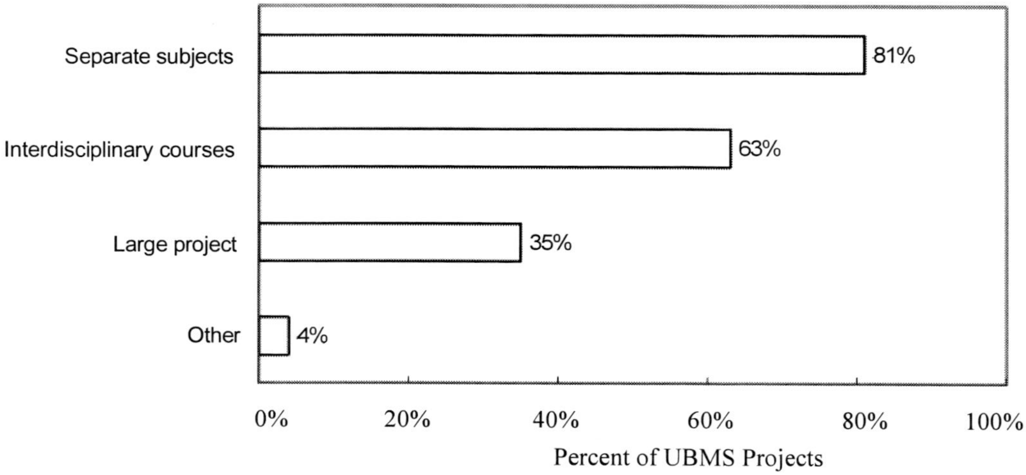

B. Primary Method Used

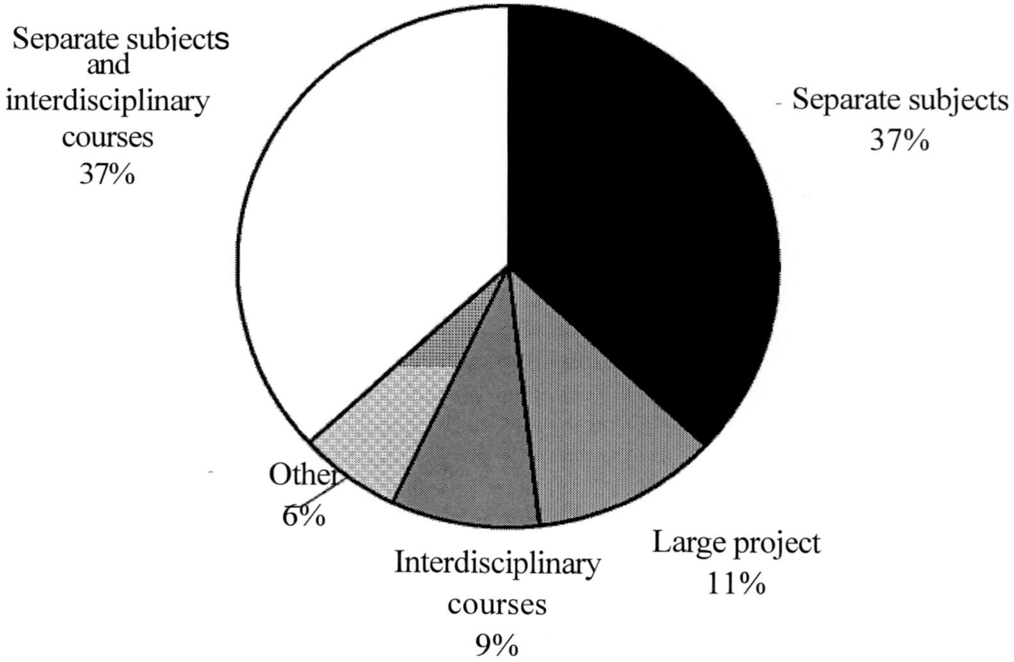

Note: Percentages may not sum to 100 percent due to rounding.
Source: 1998 survey of UBMS projects.

Exhibit II.13. Per-Capita Funding for UBMS Projects, FY 2004

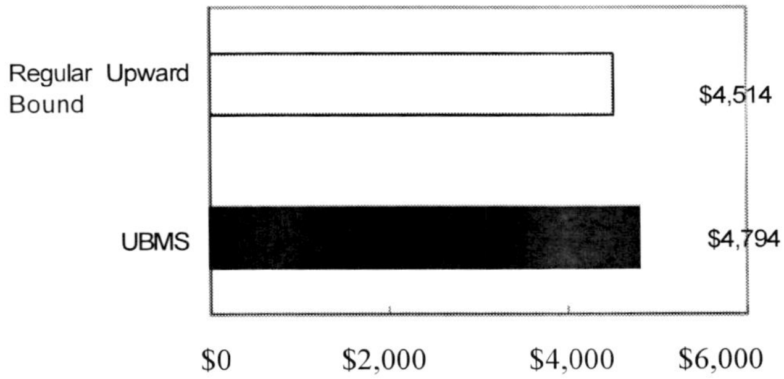

Source: U.S. Department of Education Web page (www.ed.gov/programs/triomathsci/funding.html and www.ed.gov/programs/trioupbound/funding.html accessed June 2006).

Most UBMS participants seem to stay for the entire summer. At the average project in the summer of 1994, 94 percent of participants completed all the requirements of the program. A few projects reported completion rates of 50 percent or less, but this does not necessarily indicate a high dropout rate. Students might have attended the summer program for its full length but be counted as failing to "complete all the requirements of the program"—the wording in the questionnaire—because they did not, for example, turn in all their assignments.

Some students participate for multiple summers or receive services during the academic year. As indicated earlier, about 70 percent of UBMS projects operating in summer 1994 allowed students to return from the previous summer. Among those projects, an average of 35 percent of 1994 participants had also participated during 1993.[29] Furthermore, some UBMS projects extended students' exposure to the program beyond the summer by providing services during the academic year (see Section C.4 for more details).

III. The Impacts of the Upward Bound Math-Science Program

UBMS was established to increase economically disadvantaged students' achievement in high school math and science courses, to increase the likelihood that they would major in math and science in college, and ultimately to increase their representation in math and science careers. Until now, no rigorous studies have measured the extent to which the program achieves its goals. In this chapter, we assess the effects of UBMS on important outcomes for program participants four to eight years after graduating from high school.

This chapter presents our estimates of the effects of UBMS on (1) high school preparation for college and for majoring in math and science; (2) college enrollment, highest level of college attended, college selectivity, years of college completed and highest degree earned; and (3) field of study in college. These estimates are based on information collected

[29] In summer 1998, not only did more UBMS projects (over 90 percent) allow students to return, but at those projects an average of 52 percent of students had also participated in summer 1997.

in 2002 or earlier, and another round of data collection is currently underway. Given many students in our sample may have, for example, selected a major or completed a degree between 2002 and 2004, the findings in this report should be treated as an initial assessment of the effects of the program. Our final assessment will be based on the data collected in 2004.

The evidence suggests that in addition to improving students' grades in high school, UBMS increases the likelihood of:

- Taking chemistry and physics in high school
- Enrolling in more selective four-year colleges and universities
- Completing a postsecondary degree in a math and science field

The effects of UBMS are nontrivial and in the expected direction. For example, UBMS increased the likelihood of earning a bachelor's degree in a math or science field four to eight years after high school graduation from 12 percent to 20 percent. in the next report on UBMS, we will weigh the size of the program's benefits against its costs.

Because UBMS participants in our sample were not randomly selected to participate in the program, the impact estimates presented in this chapter may suffer from selection bias. However, in designing the study, we identified several likely sources of bias and have addressed them through a combination of data collection and statistical methods described later in the chapter. Therefore, the research design is stronger than in the typical nonexperimental study, and as a result, the impact estimates should be more credible.

A. Study Design

UBMS provides intensive academic enrichment in math and science. Like regular Upward Bound projects, most UBMS projects also offer some assistance in preparing for college, such as assistance with college applications (see Chapter II for more details). The combination of intensive academic enrichment in math and science with college preparation assistance suggests that UBMS might have positive effects on the following outcomes for program participants: (1) performance in high school, especially in math and science courses; (2) postsecondary attendance, persistence, and completion; and (3) the likelihood of completing a postsecondary degree in a math and science field. Therefore, we designed the analysis to answer the following three research questions:

1. What are the effects of UBMS participation on student performance in high school overall and in math and science courses in particular?
2. What are the effects of UBMS participation on college attendance, attendance at different types of colleges and universities, years of college, and college completion?
3. What are the effects of UBMS participation on the likelihood of completing a degree in math or science?

We have designed the analysis to measure the impacts of UBMS on two important subgroups: students who had previously participated in regular Upward Bound and students who had not participated. UBMS participants who previously participated in regular Upward Bound may have received a large dose of precollege services and academic preparation before participating in UBMS. However, most of the other UBMS participants entered UBMS without having received such intensive services. It is reasonable to expect UBMS to have larger effects on the students who had not previously received intensive services: students who have already received them may have already received the boost they needed to succeed. On the other hand, UBMS participants may be better prepared to benefit from their participation if they have previously participated in intensive services. in our analysis, we compute the effects separately for students who had previously participated in regular Upward Bound and students who had not participated to assess whether the effects of UBMS depend on the amount of precollege services students have received to that point. Regular Upward Bound is just one of the other programs in which UBMS participants could have participated. However, in our analysis, regular Upward Bound deserves special consideration because many UBMS participants participate in regular Upward Bound, and because few other programs are as intensive as UBMS.[30]

It is important to clarify we have not attempted to measure the effects of UBMS versus no precollege services. The analysis was designed to measure the effects of participating in UBMS relative to what students would have participated in otherwise—which might include regular Upward Bound—*not* the effects of participating in UBMS relative to no program participation. Most UBMS participants in our sample did not participate in regular Upward Bound, and only a few other precollege programs are as intensive as Upward Bound. Therefore, most UBMS participants would have participated in less intensive precollege services if they had not participated in UBMS.

In this section, we describe the design of the analysis used to measure the impacts of UBMS. The impact analysis is based on a matched comparison group that attempts to reduce two types of bias common to many nonexperimental studies: selection bias and bias attributable to different data collection protocols for the participant and comparison groups. The strength of the impact analysis rests on three features of its design:

1. How we selected our initial samples, particularly the comparison sample
2. How we collected "baseline" (preprogram) information on the two samples
3. How we used that information to select a matched comparison group

In the remainder of this section, we describe these three features of the study and our approach to estimating the effects of UBMS on student participants.

[30] Moore (1997), Appendix A includes a list of 28 alternative math and science precollege programs. However, some other precollege programs without a specific precollege focus, such as regular Upward Bound, may share some common features with UBMS. Annual per student costs were available for 17 of the 28 programs; of these 17, only two were more expensive per student-year than UBMS.

1. Selecting the Samples

For the impact analysis, we obtained our sample of program participants from the projects themselves. In 1998, we contacted the 62 Upward Bound Math and Science Centers (MSCs) that were operating at that time and that had been operating between 1993 and 1995.[31] From these MSCs, we requested lists of the students who had participated in their program in the summer of 1993, the summer of 1994 or the summer of 1 995.[32] To reduce the costs of collecting the necessary data, we selected one out of every four of the students from these lists for our analysis sample.

A primary feature of any nonexperimental evaluation is the choice of a comparison group. Experiments yield the best comparison groups because differences in outcomes between treated and untreated cases cannot be attributed to selection bias. In the absence of an experiment, the strength of an evaluation depends on the comparability between the participant and comparison groups.

The most convenient comparison group for the impact analysis is also a compelling one— and the one we used. For the comparison group, we selected students from the evaluation of regular Upward Bound who reported that they had not participated in an MSC.[33] In doing, so, we selected a comparison group with three desirable attributes:

1. ***Like UBMS participants, comparison students applied to participate in Upward Bound.*** Therefore, both UBMS participants and comparison students revealed a high level of motivation to pursue precollege services. This provides some protection against a common source of selection bias in nonexperimental studies— bias from comparing more motivated participants to less motivated nonparticipants.
2. ***Like UBMS participants, comparison students met the federal eligibility requirements to participate in some type of Upward Bound program—regular or math and science.*** The federal eligibility requirements are the same for both regular Upward Bound and UBMS. To be included in the sample for the evaluation of regular Upward Bound—the sample from which we selected our comparison group—a student must have applied to a regular Upward Bound project and been determined eligible to participate. Therefore, both UBMS

[31] While we would have been interested in obtaining lists of students from projects that were no longer operating in 1998, we believed that it would be very difficult to obtain such lists. If MSCs that closed before 1998 operated less effective programs than those that remained open, then the results presented in this chapter may overstate the effectiveness of MSCs operating between 1993 and 1995.

[32] We received participant lists from all but one of these MSCs.

[33] If UBMS projects admitted eligible applicants on a fairly random basis, rejected applicants would probably constitute the best comparison group. However, MSCs are not required to select randomly from eligible applicants, and the evidence suggests that they do not. Case studies of 14 MSCs in the mid-1990s suggest that many MSCs impose additional eligibility requirements beyond the federal requirements, including an interest in math and science (Moore 1997). Therefore, it is unlikely that rejected applicants would have the virtues of a randomly selected control group for the purposes of this evaluation. Furthermore, the difficulty in obtaining lists of program participants from MSCs several years after they participated in UBMS suggests that many MSCs would not have been able to provide information on rejected eligible applicants.

participants and comparison students in our sample must either have come from "low-income" families (income below 150 percent of the poverty line) or potential "first- generation" families (neither parent had earned a bachelor's degree).

3. ***Like UBMS participants, many comparison students would have met project-specific eligibility requirements imposed by some MSCs.*** Moore (1997) indicates that MSCs often apply additional admissions criteria in selecting applicants—criteria that include a minimum grade point average (GPA) in math and science. in this chapter, we show that many comparison students met the same criteria: many were successfully "matched" to UBMS participants who took similar courses and earned similar grades in math and science in ninth grade.

While the regular Upward Bound sample is a useful comparison group for measuring the effects of UBMS, it is not a perfect one. Data from the 1 990s suggest that MSCs typically had more stringent minimum GPA requirements than regular Upward Bound projects; and MSC staff—who typically had recent experience with regular Upward Bound participants—reported that UBMS participants tended to be "more serious about school" than regular Upward Bound participants (Moore 1997, p. 26). Therefore, UBMS participants might fare better than regular Upward Bound participants even without any assistance from UBMS, and simple differences in mean outcomes between the UBMS participant and comparison groups may overstate the effects of UBMS due to selection bias.

To reduce selection bias, we selected a matched comparison sample from the regular Upward Bound sample. More specifically, we matched each UBMS participant to one or more regular Upward Bound sample members with similar characteristics based on data collected from student surveys and transcripts.[34] See Section 3 for more details.

2. Collecting Baseline Data

The baseline variables constructed for the impact analysis characterize members of the two samples early in high school before UBMS participants in our sample entered UBMS. These variables are critical because they allow us to account—through a combination of matching and regression adjustments—for many preexisting differences between the two groups that might otherwise bias our impact estimates. To collect the information necessary to create baseline variables, we conducted student surveys and collected high school transcripts (see Appendix A).

An important strength of the study's design is that the data collection strategy was similar for both samples. Heckman et. al. (1997) argue that different survey questionnaires for the participant and comparison samples can be an important source of bias in nonexperimental studies. In this evaluation, we developed the initial survey questionnaire for UBMS participants from the baseline survey questionnaire for the regular Upward Bound sample. Therefore, the survey questions from which we constructed baseline variables were often

[34] Very few of students in the regular Upward Bound sample participated in UBMS, and those who reported participating in UBMS were excluded from the comparison group.

identical and always similar for the two samples.[35] Furthermore, the approach to collecting and coding high school transcripts was the same for the two samples. Therefore, it is unlikely that differences in the data between the two samples have biased the impact estimates presented in this chapter.

The baseline variables for the impact analysis fall into the following three categories: (1) demographic and family characteristics, (2) participation in other precollege programs and (3) ninth-grade academic achievement in math and sciences and more generally (see Exhibit III.1). We believe that the measures of students' ninth-grade academic achievement are critical to the strength of the study. Given the findings in earlier reports, it seems entirely possible that even among students with similar demographic and family characteristics, students who participated in UBMS might have a higher academic aptitude and interest in math and science than students who participated in regular Upward Bound. Therefore, we use information on ninth-grade courses taken and ninth-grade grade point average—overall and specifically math and science— to control for differences between the two samples in early high school achievement.

Exhibit III.1. Baseline Variables

Category	Variables	Source
Demographic and family	Sex Race and ethnicity Census region Native English speaker Mother's time in the U.S. Sibling participated	Initial Surveys
Prior program participation	Sample member participated in regular Upward Bound	Initial survey and project records High school transcripts
Ninth-grade achievement	GPA in math and science GPA in other subjects Math course taken Science course taken	

Note: The initial surveys were conducted in 1999 for UBMS participants and in 1992-94 for comparison students in the regular Upward Bound sample. To identify UBMS participants who had previously participated in regular Upward Bound, we used responses to the 1999 initial survey; to identify comparison students who participated in regular Upward Bound, we used participation information provided by projects. We collected high school transcripts in 2000 and 2003 for UBMS participants and in 2000 and earlier years for comparison students.

[35] Two differences in the two surveys were addressed in constructing baseline variables for the MSC impact analysis: (1) the two survey questionnaires were different, and (2) the surveys were conducted at different times— after high school for the participant group and early in high school for the comparison group. To address the differences in the survey questionnaires, we used data from survey questions that are either identical or almost identical. To address the difference in the timing of the surveys, we used data from survey questions only when the timing of the survey was unlikely to affect the answer to the question.

Exhibit III.2. Summary Statistics from the Baseline Variables (Percentage unless otherwise noted)

Characteristic	UBMS Participants	Regular Upward Bound Sample	Matched Comparison Group
Participated in Regular Upward Bound	18	55***	18
Female	59	72***	59
Race and Ethnicity			
African American	37	37	37
White	25	34***	30
Hispanic	18	20	16
Other race	20	9***	17
Region			
Northeast	11	9	12
Midwest	23	19**	28
South	40	45**	35
West	25	26	24
Entry to High School			
1991-92	32	25***	28
1992-93	39	49***	37
1993-94	29	26	35
Other Characteristics			
Native English speaker	80	87***	86**
Mother in U.S. most of her life	79	87***	83*
Siblings in Upward Bound	11	11	12
Ninth-Grade Math Course			
Lower than algebra	16	32***	14
Algebra	55	53	59
More than algebra	29	14***	27
Ninth-Grade Science Course			
Biology, chemistry, or physics	37	26***	37
Ninth-Grade GPA (mean)			
Math and science	2.69	1.65***	2.71
Other subjects	3.24	2.64***	3.25

*/**/*** Significantly different from UBMS participants at the 0.10 / 0.05 / 0.01 level.

3. Selecting a Matched Comparison Sample for the Impact Analysis

The data collected on UBMS participants and regular Upward Bound sample members are useful in identifying many similarities between the two groups—and some differences as well (see Exhibit III.2). In both groups, 37 percent of the students were African American and 11 percent had a sibling who had participated in Upward Bound. The two groups were similar with respect to many other characteristics as well, including the percentage taking algebra in ninth grade. However, the two groups exhibit differences on several dimensions. For example, UBMS participants were more likely to be male and tended to have higher grades than regular Upward Bound sample members.

To address possible selection bias, we selected a matched comparison group from the regular Upward Bound sample using propensity score matching methods.[36] The goal in matching was to select a matched comparison sample from the regular Upward Bound sample such that the distributions of the baseline variables for the UBMS participant sample and the matched comparison sample were similar.[37] Matching was conducted separately for sample members who had previously participated in regular Upward Bound and for those that had not:

- UBMS participants who had previously participated in regular Upward Bound were matched to members of the treatment group in the evaluation of regular Upward Bound.
- UBMS participants who had not previously participated in regular Upward Bound were matched to regular Upward Bound control group members who did not participate in regular Upward Bound.

Regular Upward Bound sample members that were matched to at least one UBMS participant were included in the matched comparison sample.[38,39]

Using matching procedures, we were able to select a matched comparison group that is highly similar to the sample of UBMS participants on many dimensions (see the last column of Exhibit III.2). Only two of the differences between the groups are statistically significant—the difference in the percentage of sample members who are native English speakers and the

36 Many studies adjust for baseline differences of these types using standard covariance adjustments—that is, by controlling for these variables in a regression analysis. However, the differences between the UBMS participant sample and the regular Upward Bound sample are too large to expect covariance adjustment to be reliable. Regression adjustments are likely to be unreliable if the means of the propensity scores are more than half a standard deviation apart (Rubin 2002). For the participant sample, the mean and standard deviation of the propensity score are 0.58 and 0.22, respectively; for the comparison sample, the mean and standard deviation of the propensity score are 0.25 and 0.23, respectively. Therefore, the difference in mean propensity scores is more than one standard deviation, and regression adjustments alone are likely to be unreliable.

37 More specifically, the goal is to ensure that the distributions of the baseline variables for the UBMS participant sample and the matched comparison sample are similar enough such that covariance adjustments will produce reliable impact estimates.

38 The propensity score matching and the impact analysis are restricted to sample members who entered high school between 1991 and 1993. While some UBMS participant sample members entered high school before 1991 and after 1993, relatively few comparison sample members did so. Furthermore, high school cohort is related to the likelihood of participating in the program and to the outcomes of interest because earlier cohorts had more time to enter college and select a field of study by the time they were interviewed for the evaluation. Therefore, to protect the internal validity of the study, we focused the analysis on students who entered high school between 1991 and 1993.

39 UBMS participants could be matched to more than one regular Upward Bound sample member, and regular Upward Bound sample members could be matched to more than one UBMS participant. To be matched, a pair of students must satisfy the following condition: the difference between matched students in the log odds of the propensity scores was less than

percentage whose mother has lived in the United States for all or almost all of her life. Furthermore, given the number of baseline variables, two is a small number of significant differences: we would expect about two significant differences even if the differences between the two groups were purely random. While there is no guarantee that matching removed all unmeasured differences between the two samples, matching removed differences on a broad range of baseline variables—differences that might otherwise bias the impact estimates.

4. Estimating the Impacts of UBMS Participation

To measure the effects of UBMS participation on participating students, we used a regression-based approach that allows us to (1) adjust for the small remaining differences between the UBMS participant sample and the matched comparison group and (2) increase the precision of our impact estimates. The regression models yield estimates of the effect of UBMS on students who participated in the program.[40]

We estimated the effects of UBMS for the entire sample and separately for selected subgroups of students. The effects of UBMS may depend on the amount of other precollege services received. Therefore, we present separate impact estimates for those who participated in regular Upward Bound and those who did not participate. Furthermore, we present separate estimates by sex and race since achievement in math and science often varies along these dimensions, and the effects of UBMS may vary along the same dimensions.[41,42]

0.20 times the standard deviation of the log odds. Smith and Todd (2003) refer to this type of matching as "radius matching."

[40] We regressed each outcome on a set of control variables and an indicator of whether the student participated in UBMS. The control variables included the variables used in selecting the matched comparison group: prior participation in regular Upward Bound, siblings in Upward Bound, sex, race, ethnicity, mother's native language and immigrant status, high school cohort, region of the country, and several variables describing academic achievement in ninth grade, including GPA—separately for math and science courses and for other courses—math course taken, and science course taken. For continuous variables, such as number of college credits, we estimated linear regression models; for categorical outcomes, such as whether the sample member pursued postsecondary studies in math or science, we estimated logistic regression models or "logit" models. In estimating standard errors, we accounted for clustering by project and used the Taylor series linearization methods employed by the SUDAAN statistical analysis software.

[41] Other reports from the National Evaluation of Upward Bound have shown that the effects of regular Upward Bound vary with the educational expectations of student applicants (see Myers et al. 2004 and Myers and Schirm 1999). In particular, regular Upward Bound greatly increased the likelihood of attending a four-year college or university for students who did not expect to earn a bachelor's degree when they applied to participate in regular Upward Bound. Because we first interviewed UBMS participants after they finished participating in UBMS, we lack information on their educational expectations when they applied to participate in UBMS. Therefore, we cannot assess whether the effects of UBMS vary with students' educational expectations before participating.

[42] In the national evaluation, we have estimated the effects of regular Upward Bound separately by eligibility category (low-income, first-generation, or both). However, student data was provided by project directors when they submitted students to Mathematica for random assignment. Because we did not conduct random assignment for the evaluation of UBMS, we lack data on eligibility for UBMS participants. While we did obtain rosters of prior participants

B. Analysis and Findings

In this section, we present the estimated effects of UBMS on students' outcomes. Consistent with program objectives, participation in UBMS seems to promote postsecondary study in math and science. Specifically, our estimates suggest that UBMS participation (1) raises grades in high school and course-taking in chemistry and physics, (2) increases enrollment in more selective four-year institutions and the number of years of college completed, and (3) raises the likelihood of both pursuing postsecondary studies and completing a four-year degree in math or science. In contrast, UBMS does not affect course-taking in advanced math in high school.

1. The Effect of UBMS on High School Outcomes

The UBMS program is designed to strengthen participants' math and science skills. MPR conducted follow-up student surveys and collected high school transcripts to determine whether UBMS participation increased the amount of course work in math and science as well as total high school credits, high school graduation and GPA in math and science and overall.[43]

UBMS raised the likelihood of taking upper-level high school courses in science but not in math (see Exhibit III.3). Our estimates indicate that UBMS increased the likelihood of taking upper-level science courses in high school, raising the percentage of students taking chemistry from 78 to 88 percent and raising the percentage of students taking physics from 43 to 58 percent.[44] The estimated effects of UBMS on the likelihood of taking algebra II, trigonometry, precalculus and calculus are statistically insignificant.

It is not clear why UBMS affected course-taking in the sciences but not in math. One possibility involves differences between the type of instruction that MSCs offer in math and the type of instruction they offer in sciences. Almost two-thirds of MSCs reported offering interdisciplinary classes, and about a third of them offered instruction that revolved around a large project or experiment (see Chapter II). However, findings from Moore (1997) suggest that in MSCs, laboratories and hands-on instruction were much more common in science instruction than in math instruction, and that MSC instructors used math to help student prepare for careers in science and technology (see Moore 1997, pp. 41-42). It is possible that the combination of hands-on instruction offered only in science with the science orientation of MSC instructors is responsible for UBMS's effects on science course-taking.

Additional estimates suggest UBMS raises overall achievement and educational attainment in high school (see Appendix B). UBMS had a positive effect on high school grades, increasing the average GPA in math courses from 2.7 to 2.8, the average GPA in science courses from 2.7 to 2.9 and the average GPA overall. It also raised the graduation rate from 96 to 99 percent.

---

from UBMS projects in 1998, we did not request data on eligibility because we did not believe that projects could easily provide them for students who participated three or more years ago.

[43] With the exception of the outcome variables describing high school completion, which were constructed from student survey responses, the high school outcomes were constructed using data from high school transcripts.

UBMS appears to have had larger effects on grades and science course work for Hispanics than for African Americans (see Appendix B). For example, UBMS raised the average GPA in math courses from 2.4 to 2.5 for African Americans and from 2.4 to 2.7 for Hispanics. In addition, UBMS raised the likelihood of taking chemistry and physics by 7 percentage points each for African Americans and by 17 and 27 percentage points, respectively, for Hispanics.[45] Comparisons of males to females and students who had participated in regular Upward Bound to other students revealed few significant differences.

2. The Effect of UBMS on Postsecondary Attendance, Persistence, and Completion

While UBMS focuses on preparing students to major in math and science and to complete a degree in a math and science field, a person must enroll in college before choosing a major and must complete college to earn a degree in a math and science field. Even if UBMS had little effect on students' choice of major, UBMS might be a cost-effective strategy to increase college enrollment and completion for disadvantaged students. Therefore, we assess whether UBMS promotes postsecondary attendance, persistence, and completion before examining its effects on college major.

Exhibit III.3. Math and Science Courses Taken in High School (Percentage of Students)

	UBMS Participants	Matched Comparison Group	Impact
Math Courses			
Algebra	97	97	0
Geometry	86	86	0
Algebra II	73	74	-1
Trigonometry	35	30	5
Analysis or Precalculus	36	31	5
Calculus	26	24	2
Science Courses			
Biology	97	98	-1
Chemistry	88	78	10***
Physics	58	43	15***

*/**/*** Statistically significant at the 0.1 /0.05 / 0.01 level.

Notes: Impact may not exactly equal the difference between UBMS participants and the matched comparison group due to rounding. See Table B. 1 in Appendix B for more details and additional high school outcomes and Tables B.2 through B.4 for estimates for subgroups of UBMS participants.

[44] These increases in part reflect increases in the total number of courses taken in high school: UBMS raised the average number of high school credits earned from 25 to 26.

[45] For African Americans, the estimated effect on the likelihood of taking physics was not statistically significant.

Exhibit III.4. Postsecondary Attendance, Highest Level Attended and College Selectivity (Percentage of Students)

	UBMS Participants	Matched Comparison Group	Impact
Postsecondary Attendance	95	90	5***
Highest Level Attended			
Four-year college or university	82	71	1 1***
Two-year college	11	16	-5***
Vocational institution	2	4	-2
Most Selective Four-Year institution			
More selective	33	23	11
Less selective	48	48	0

*/**/*** Statistically significant at the 0.1 /0.05 / 0.01 level.

Notes: Impact may not exactly equal the difference between UBMS Participants and the Matched Comparison Group due to rounding. "Most Selective College or University" refers to four-year colleges and universities only, and some four-year institutions were not classified in *Barron's Profiles of American Colleges* (2003). See Table B.5 in Appendix B for more details and additional postsecondary outcomes and Tables B.6 through B.8 for estimates for subgroups of UBMS participants.

The evidence suggests that UBMS participation increases the likelihood of attending a postsecondary institution. Estimates for the matched comparison group suggest that 90 percent of UBMS participants would have attended a postsecondary institution if they had not participated in UBMS (see Exhibit III.4). This indicates that UBMS participants are much more likely to attend college than the average low-income high school student, and a 90-percent attendance rate leaves little room for improvement. Despite this, the evidence suggests that UBMS participation increases the likelihood of attending a postsecondary institution from 90 to 95 percent, although as shown in Appendix C, this result is sensitive to how postsecondary attendance is measured and, specifically, to whether college transcripts are used to verify attendance.[46]

[46] This 95-percent figure suggests that almost all UBMS participants attend a postsecondary institution in the first few years after high school. In contrast, findings from a recent analysis of performance data submitted by UBMS projects seem to suggest a much lower rate of postsecondary attendance. Analyzing the performance data, Curtin and Cahalan (2004) find that only 68 percent of UBMS participants who graduated from high school in 1999- 2000 were reported as attending a postsecondary institution in 2000-01. The difference between this figure and the enrollment rate presented in this report reflects substantial differences in how the two estimates were derived. According to information provided to us by ED, the estimate in Curtin and Cahalan (2004) indicates the percentage of former program participants for whom UBMS project directors could confirm postsecondary attendance. The estimate was not intended to measure the postsecondary enrollment rate per se, and it understates the true rate for UBMS participants for two related reasons. First, the estimate includes in the denominator but not the numerator UBMS participants from projects that were no longer funded and, therefore, did not submit performance data. Second, it excludes from the numerator about

The evidence also suggests that UBMS participation increases the likelihood of attending a four-year college or university from 71 percent to 82 percent. The increase in four-year attendance can be attributed to increased attendance at more selective schools as UBMS raises the likelihood of attending such a school from 23 to 33 percent.[47] It appears that some of this increase in four-year college enrollment can be attributed to students who would otherwise have attended only two-year colleges: UBMS reduced the likelihood of enrolling in a two-year college but not in a four-year college or university from 16 to 11 percent.

UBMS increased the number of years that students spend in four-year colleges and universities without increasing the completion rate at these schools. Our estimates suggest UBMS increased the number of years enrolled in a four-year institution from 2.4 to 2.9 (see Exhibit III.5). While UBMS increased the amount of time participants spent in college, it had no effect on the likelihood of their completing a bachelor's degree as of the time we last interviewed them in 2002. Given that almost half of UBMS participants in our sample (47 percent) were attending four-year colleges and universities when we last interviewed them, it is too early to reach firm conclusions about the effects of UBMS on the number of years of college attended and on college completion. Our final report on the Upward Bound Math-Science Program will be based on data collected in 2004.

The average effects of UBMS mask some interesting differences between men and women. For both men and women, UBMS increased the percentage attending a selective four-year institution—from 23 to 34 percent for men and from 21 to 32 percent for women. However, UBMS only affected overall four-year enrollment for women. For women, UBMS increased the likelihood of attending a four-year institution from 68 to 82 percent and reduced the likelihood of attending a two-year institution (without also attending a four-year institution) from 18 percent to 10 percent; for men, both of these effects were statistically insignificant.

The effects on college completion also differ between men and women. For women, the positive effect of UBMS on enrollment in four-year institutions—along with positive effects on credits earned in these institutions—translates into higher graduation rates from four-year institutions. UBMS raised the percentage of women earning a bachelor's degree from 32 to 40 percent. For men, the effect of UBMS on the percentage completing a bachelor's degree is statistically insignificant; however, UBMS increased the percentage completing an associate's degree from 4 to 8 percent.

In the last report from the national evaluation (Myers et al. 2004), we presented two sets of estimates for postsecondary attendance and highest level attended—the first based entirely

---

one-fifth of the participants who are included in the denominator because UBMS project staff were unable to confirm college attendance for these former participants. For our estimate, in contrast, we exclude from both the numerator and denominator the UBMS participants in our sample for whom we are missing data due to survey nonresponse and, then, use standard weighting techniques to adjust for the missing observations.

[47] If a school was rated as "most competitive," "highly competitive," or "very competitive," then we classified the school as "more selective." If a school was rated as "competitive," "less competitive," "noncompetitive," or "special," or the school was either excluded from Barron's or was included but not rated, then we classified the school as "less selective." Students were then classified as "more selective" school if they attended one or more of the "more selective" institutions, and "less selective" if they attended one or more of the "less selective" institutions and did not attend a "more selective" institution.

on the information reported by the student, and the second that required verification of enrollment from the institution. In computing the second set of estimates, we made the strong assumption that a student did not attend the school they reported to us unless we received verification from the school in the form of a college transcript (or a reason for not providing one that clearly indicated that the student had attended the school). However, additional analysis of the data suggests that in most cases for which we did not receive a transcript, there is no reason to doubt that the students actually attended the schools they reported. Schools provided a variety of reasons for not providing transcripts, including confidentiality considerations and money that the student owed to the school. Therefore, for all outcome variables except years of college, which are based on information from college transcripts, the estimates presented in this chapter are based on the information reported by the students themselves. However, we conducted a sensitivity analysis to see whether requiring verification of enrollment affected our impact estimates, and in most cases, it did not. See Appendix C for more details about the sensitivity analysis.

Exhibit III.5. Years of College and Degree Completion

	UBMS Participants	Matched Comparison Group	Impact
Years of College			
Four-year college or university	2.9	2.4	0.4**
Two-year college	0.4	0.4	0.0
Degree Completion (percent)			
Bachelor's degree	35	33	2
Associate's degree	7	7	-1

*/**/*** Statistically significant at the 0.1 /0.05 / 0.01 level.

Notes: The comparison group estimates and impact estimates are regression-adjusted (see Chapter III, Section A.4). Impact may not exactly equal the difference between UBMS Participants and the Matched Comparison Group due to rounding. Years of Postsecondary Education were computed by dividing the number of credits earned by 30. See Table B.5 in Appendix B for more details and additional postsecondary outcomes and Tables B.6 through B.8 for estimates for different subgroups of UBMS participants.

3. The Effect of UBMS on Postsecondary Field of Study

The primary objective of UBMS is to prepare students for postsecondary studies in math and science. Using information reported by students in 2002, we conducted an analysis to examine whether UBMS participation increases the likelihood of pursuing postsecondary studies in math and science or the likelihood of earning a postsecondary degree in a math or science fields. To determine whether a sample member had pursued postsecondary studies in math and science, we asked sample members in the 2002 survey for their "most recent or intended field of study," and we classified their responses according to the same classification system used in the National Science Foundation's Scientists and Engineers Statistical Data

System (SESTAT).[48] Furthermore, for the analysis, we separated the social sciences from other math and science fields (which we refer to as "math or science") because the objectives of the program are more closely tied to the latter than the former.

The evidence suggests that UBMS participation encourages students to pursue postsecondary studies in math or science and also in the social sciences. UBMS increased the likelihood of majoring or intending to major in math or science from 23 to 33 percent overall and from 18 to 28 percent if we focus on majors at four-year colleges and universities (see Exhibit III.6).[49] The effects are also positive but smaller in magnitude for the social sciences: UBMS increased the likelihood of pursuing postsecondary studies in the social sciences from 7 to 11 percent overall and from 7 to 10 percent if we focus on majors at four-year institutions.

The evidence also suggests that UBMS participation may encourage students to complete postsecondary degrees in math or science. However, with almost half of participants still in college, it is too early to reach final conclusions about UBMS's effects on college completion overall or in specific fields. As of the 2002 survey, the effect of UBMS on the likelihood of earning a degree or certificate in math and science was statistically insignificant. However, if we focus on four-year institutions, UBMS altered the types of fields in which students earned degrees. More specifically, it raised the likelihood of earning a bachelor's degree in math or science from 6 to 12 percent and reduced the likelihood of earning a bachelor's degree outside of math or science or the social sciences from 20 to 14 percent.

The effect of UBMS on the likelihood of majoring in math and science was larger for men than for women. While UBMS increased the likelihood of pursuing postsecondary studies in math and science from 38 to 58 percent for men, the estimated effect for women is much smaller and statistically insignificant. In contrast, there is no evidence that the effect of UBMS on the likelihood of completing a math or science degree differs between the two groups.

C. Interpretation of the Findings

The estimated effects of UBMS paint a fairly consistent picture of UBMS helping its participants onto an educational path that could lead to careers in math and science. Our findings suggest that UBMS participation improves student outcomes in high school and college, and— consistent with the objectives of the program—increases the odds of completing a college degree in math and science.

While we took several steps to reduce selection bias, it is certainly possible that the true effects of participating in UBMS are smaller than our estimates suggest. As reported earlier in

[48] The fields classified as science and engineering were biological sciences, computer science, engineering, mathematics, physical sciences, and technical fields. The fields classified as non-science and engineering were agriculture, arts, business, education, clerical or legal assistance, communications, health-related fields, humanities, trade and industry, protective services and consumer or personal services. A small number of fields reported by sample members could not be classified as either science and engineering or nonscience and engineering. (For more details about NSF's classification of fields of study, see www.nsf.gov/sbe/srs/nsf99337/pdf/appa.pdf.)

[49] For convenience, we use the terms "major" and "field of study" interchangeably.

the chapter, UBMS participants may be more serious about school than regular Upward Bound participants on average. This difference is reflected in our data: UBMS participants in our sample had higher GPAs and took more advanced math and science courses in ninth grade than members of the regular Upward Bound sample. We accounted for these differences by selecting a matched comparison group that resembled the UBMS participant sample in both grades earned and course-taking in ninth grade. However, it is possible that despite earning similar grades and taking similar courses early in high school, the UBMS participant sample is somewhat more serious about school, more serious about math and science, or is different from the matched comparison group in some other way that would lead our analysis to overstate the effects of UBMS.

Exhibit III.6. Field of Study (Percentage of students)

	UBMS Participants	Matched Comparison Group	Impact
All Postsecondary institutions			
Postsecondary Studies			
Math or science	33	23	10***
Social science	11	7	4*
Other	42	51	-9***
Postsecondary Studies Completed			
Math or science	15	12	3
Social science	8	4	3*
Other	21	28	-7**
Four-Year Colleges and Universities			
Postsecondary Studies			
Math or science	28	18	10***
Social science	10	7	4*
Other	36	39	-3
Postsecondary Studies Completed			
Math or science	12	6	6***
Social science	6	4	3
Other	14	20	-6**

*/**/*** Statistically significant at the 0.1 /0.05 / 0.01 level.

Notes: Impact may not exactly equal the difference between UBMS Participants and the Matched Comparison Group due to rounding. See Table B.9 in Appendix B for more details and additional outcomes and Tables B. 10 through B. 12 for estimates for subgroups of UBMS participants.

While we cannot measure the extent of selection bias, some informed speculation is helpful in interpreting the impact estimates. As we indicated earlier, students in both groups—the UBMS participant sample and the matched comparison group—exhibited some motivation to improve academically by applying to participate in Upward Bound. Therefore, motivational differences between the two groups are likely to be small and unlikely to bias the impact estimates. However, the impact estimates would overstate the true impacts if:

- ***Members of UBMS participant sample were higher achievers than members of matched comparison group.*** by the end of high school and after students in the UBMS sample had participated in UBMS, the average GPA was slightly but significantly higher for UBMS participants (3.14) than for matched comparison students (3.06). Even if the entire difference were attributable to selection bias instead of the effects of the program, the average student in each group was a B student. Therefore, if the UBMS participant sample contains higher achieving students than the matched comparison group, the difference seems to be small.
- ***Members of the UBMS participant sample were better at math and science than members of the matched comparison group.*** If students in the UBMS participant sample were better at math and science than students in the matched comparison group, it is not reflected in their math and science course-taking or grades during ninth grade: the two groups had similar grades in math and science, were equally likely to take algebra or geometry and were equally likely to take biology, chemistry or physics. While we did not explicitly examine whether UBMS participants were more likely to enroll in the advanced sections of courses before entering UBMS, other findings suggest that UBMS participants were no more likely to have taken Advanced Placement classes in math or science by the time they finished high school. Finally, focus groups conducted in 1996 suggest that UBMS participants do not view themselves as particular strong in math and science (Moore 1997, p. 28). Therefore, the information we have collected provides no reason to believe that when they entered UBMS, the participant sample was better at math and science than the matched comparison group.
- ***Members of the UBMS participant group were more interested in math and science than members of the matched comparison group.*** None of the information that we extracted from student transcripts would suggest that the UBMS participant sample had greater interest in math and science. It is possible UBMS participants had greater interest in careers in math and science that simply was not reflected in their high school course-taking or grades early in high school. in 1996 focus groups, many UBMS participants expressed interest in pursuing careers in scientific fields, such as engineering, medicine, and nursing (Moore 1997, p. 28). However, many of these students indicated that their career interests had developed just that summer, and the expression of those interests could have been influenced by the fact that the focus groups were conducted on site at projects that emphasized math and science.

We suspect that the impacts most vulnerable to selection bias are those that are most closely related to a person's interest in pursuing careers in a math or science field. To gauge students' interest in math and science early in high school would probably require conducting assessments or survey interviews at that time. However, we first interviewed members of the UBMS participant sample after they had completed high school. The possibility that UBMS participants might have had greater interest in pursuing a career in science than matched comparison students raises the question of whether our estimates overstate the effects of

UBMS on the outcomes that are most closely related to one's career interests, such as majoring in math or science in college.

Therefore, while the findings in this report are promising, a note of caution is appropriate. We speculate that the selection bias is likely to be largest for outcome variables most closely tied one's interest in pursuing math and science careers, but it is not possible to measure the selection bias. While we took several steps to reduce selection bias, the estimated effects of UBMS may overstate the true effects of the program.

References

Cahalan, Maggie, Tim Silva, Justin Humphrey, Melissa Thomas and Kusuma Cunningham. *Implementation of the Talent Search Program, Past and Present. Final Report from Phase I of the National Evaluation.* Report prepared for the U.S. Department of Education, Planning and Evaluation Service. Washington, D.C.: Mathematica Policy Research, Inc., October 2002.

Curtin, Thomas R., and Margaret W. Cahalan. *A Profile of the Upward Bound Math-Science Program: 2000-2001.* Washington, D.C.: U.S. Department of Education, Office of Postsecondary Education, December 2004.

Fasciano, Nancy J., and Jon E. Jacobson. "Grantee Survey Report" in *A 1990s View of Upward Bound: Programs Offered, Students Served, and Operational Issues*, a monograph prepared by Mathematica Policy Research, Inc., for the U.S. Department of Education, Office of the Under Secretary, as part of the National Evaluation of Upward Bound. Washington, D.C.: May 1997.

Moore, Mary T. *A 1990s View of Upward Bound: Programs Offered, Students Served, and Operational Issues*. In the monograph of the same name, prepared by Mathematica Policy Research, Inc., for the U.S. Department of Education, Office of the Under Secretary, as part of the National Evaluation of Upward Bound. Washington, D.C., May 1997.

Moore, Mary T. *Developing Math and Science Skills Among Disadvantaged Youth: A Review of the Upward Bound Precollege Math/Science Centers.* Report prepared for the U.S. Department of Education, Planning and Evaluation Service. Washington, D.C.: Mathematica Policy Research, Inc., September 1997.

Myers, David. *Measuring the Impact of the Upward Bound Math/Science Initiative: Feasibility of an Evaluation.* Report for the Department of Education. Washington, D.C.: Mathematica Policy Research, Inc., May 7, 1997.

Myers, David, Rob Olsen, and Neil Señor. The Impacts of Upward Bound: High School Experiences and College Access. Presentation at Annual Conference, Council for Opportunity in Education, Washington, D.C., September 2002.

Myers, David, Robert Olsen, Neil Señor, Julie Young, and Christina Tuttle. *The Impacts of Regular Upward Bound: Results from the Third Follow-up Data Collection.* Report prepared for the U.S. Department of Education, Policy and Program Studies Service. Washington, D.C.: Mathematica Policy Research, Inc., April 2004.

Myers, David, and Allen Schirm. *The Impacts of Upward Bound: Final Report for Phase I of the National Evaluation.* Prepared for the U.S. Department of Education, Planning and Evaluation Service. Washington, D.C.: Mathematica Policy Research, Inc., April 1999.

National Science Foundation. "Table C-11. Employed U.S. Scientists and Engineers, by Level and Broad Field of Highest Degree Attained, Race/Ethnicity, and Years Since Degree: 1995." Available at http://srsstats.sbe.nsf.gov/preformatted-tables/1995/tables/tbC11.pdf, accessed June 2006.

Rubin, Donald B. "Using Propensity Scores to Help Design Observational Studies: Application to the Tobacco Litigation." *Health Services and Outcomes Research Methodology*, 2, pp. 169-188, 2002.

U.S. Census Bureau. *Statistical Abstract of the United States: 2001*. Washington, D.C., November 2001.

U.S. Census Bureau. "DP-1. General Population and Housing Characteristics: 1990." Online data table from 1990 Census. (http://factfinder.census.gov/servlet, accessed June 2006).

U.S. Census Bureau. "QT-P1D. Age and Sex of the Black Population: 1990." Online data table from 1990 Census (http://factfinder.census.gov/servlet, accessed June 2006).

U.S. Department of Education, Office of Educational Research and Improvement, National Center for Education Statistics. *NAEP 1999 Trends in Academic Progress: Three Decades of Student Performance*. NCES 2000-469, by J.R. Campbell, C.M. Hombo, and J. Mazzeo. Washington, D.C., 2000.

U.S. Department of Education, Office of Educational Research and Improvement, National Center for Education Statistics. *Digest of Education Statistics 1998*. NCES 1999-036. Washington, D.C., 1999.

U.S. Department of Education, Office of Educational Research and Improvement, National Center for Education Statistics. *Confronting the Odds: Students At Risk and the Pipeline to Higher Education*. NCES 98-094. Washington, D.C., 1997.

U.S. Department of Education, Office of Educational Research and Improvement, National Center for Education Statistics. *The Condition of Education 1996*. NCES 96-304. Washington, D.C., 1996a.

U.S. Department of Education, Office of Educational Research and Improvement, National Center for Education Statistics. *The Condition of Education 1994*. NCES 94-149. Washington, D.C., 1994.

U.S. Department of Education, Office of Educational Research and Improvement, National Center for Education Statistics. *National Education Longitudinal Study of 1988. High School Seniors' instructional Experiences in Science and Mathematics*. NCES 95-278. Washington, D.C., 1996b.

APPENDIX A
DATA COLLECTION

This appendix describes and assesses the procedures for collecting the data that we used to construct student outcome measures for the impact analysis presented in Chapter III. These data come from two different sources:

1. The fourth follow-up survey of students
2. Secondary and postsecondary transcripts

This appendix focuses on procedures for obtaining completed interviews in the fourth follow-up survey and for collecting academic transcripts.

A. Fourth Follow-Up Survey of Students

The fourth follow-up survey was conducted between April 2001 and December 2002. It was designed to collect information on secondary and postsecondary educational outcomes approximately five to seven years after scheduled completion of high school.

1. Data Collection Modes

One week before we began interviewing, we sent a letter to all study participants. The letter indicated that we would call them to complete an interview for an important study, and it encouraged them to participate. In addition, the letter indicated that we would pay them $10 for completing the interview.

Most interviews were administered using computer-assisted telephone interviewing (CATI). CATI interviews took about 30 minutes to complete. When a CATI interview was not possible, we attempted to obtain a completed questionnaire through the mail. Study participants were also offered the option of completing the survey on the Web. In June 2001, questionnaires were mailed to study participants that could not be reached by telephone. Three additional follow up mailings were conducted after the first mailing, with the last set of questionnaires being sent out in January 2002.

2. Locating

Throughout the data collection period, locating staff used services such as LexisNexis and Internet databases to obtain updated addresses and phone numbers for study participants that were difficult to reach.

3. Incentives

Financial incentives for survey completion were used to obtain a high response rate. Study participants were offered a $10 incentive for participating in the survey. Incentive checks were mailed after the sample member completed the interview.

4. Response Rates

The eligible sample consisted of 1,759 UBMS participants and 2,830 sample members from the evaluation of regular Upward Bound. (See Chapter III for more details on the samples.) We obtained completed interviews for 1,425 UBMS participants and 2,146 regular Upward Bound sample members for response rates of 81 percent and 76 percent, respectively (see Table A. 1).

Table A.1. Fourth Follow-up Survey of Students

	UBMS Sample	Regular UB Sample	Full Sample
Completed Interview	1,425	2,146	3,571
Eligible Nonrespondent	334	684	1,018
Ineligible - Deceased	7	14	21
Total	1,766	2,844	4,610

B. Transcript Data Collection

Secondary and postsecondary transcripts were collected between July 2002 and March 2003. Academic transcripts provided the primary source of information on postsecondary achievement. Transcript requests were made from institutions that were reported by sample members in the fourth follow-up survey of students and in earlier surveys.

1. Preparation for Requesting Transcripts

Information about students' secondary and postsecondary enrollment was primarily obtained from follow-up interviews. Students reported the secondary and postsecondary institutions that they had attended. Secondary transcripts were only requested from UBMS sample members selected for the impact analysis; postsecondary transcripts were requested from all sample members—both UBMS and regular Upward Bound—who reported or confirmed having attended a particular postsecondary institution.[50]

To obtain mailing addresses for the schools that were attended by sample members, we matched schools that were reported by survey respondents to directories of secondary and postsecondary schools maintained by the U.S. Department of Education. Secondary schools were matched to the Common Core of Data (CCD); postsecondary schools were matched to the integrated Postsecondary Education Data System (IPEDS).[51]

2. Procedures for Requesting Transcripts

Each school was sent a transcript request packet that included:

- A letter, printed on Department of Education letterhead, which explained the

[50] We did not collect secondary transcripts for regular Upward Bound sample members because we had already collected these transcripts for a large percentage of the sample in previous waves of data collection.

[51] Students were asked to provide the name and state of each secondary and postsecondary school they attended, but sometimes misspellings or incomplete information resulted in some invalid requests for student transcripts as schools were matched with an incorrect address and transcripts were requested from the wrong school. When a school indicated that they could not fill a request because they had no record of the student whose transcript we requested, it was

purpose of the study and the reason we were requesting transcripts.

- A statement of Authorization and Confidentiality, which cited the *Family Educational Rights and Privacy Act* and included questions and answers regarding consent and confidentiality.
- A transcript checklist of all the materials that we requested from the school, including student transcripts, a course catalog, grade descriptions and a transcript reimbursement form, which would indicate the reimbursement that the school required for providing the requested transcripts.
- A postage-paid business reply envelope for sending the transcripts.
- A disclosure notice to be placed in each student's file, indicating that a copy of his or her transcript was released to Mathematica Policy Research as an agent to the U.S. Department of Education.

3. Follow-Up Procedures

For schools that did not respond to our initial request for transcripts, we mailed another request for student transcripts. These mailings were done periodically as we tracked the schools that had not yet sent the requested transcripts and corrected requests that contained errors.

As the targeted end date for collecting transcripts approached, interviewers started calling schools directly to inquire about the status of our requests. Many schools responded to these calls by faxing us the requested transcripts. When the school indicated that they could not provide one or more of the requested transcripts, the interviewer completed a problem sheet indicating the reason. The reason generally fell into one of the following categories:

- ***The student was never enrolled at the school according to the school's records.*** When this occurred, our first response was to call the school and provide more information on the student (e.g., provide or verify date of birth and dates of attendance) to see if a transcript could be located with additional information. In many cases, the school was able to locate and provide transcripts once additional information was provided. In other cases, the school provided some information that helped us determine where we might obtain the needed transcripts.[52] If the school had no record of the student having ever attended and we were unable to obtain additional information, we marked the case as an invalid request.
- ***Transcripts were held by the school district.*** Some schools only held the transcripts of currently enrolled students and all other transcripts were sent to the school district. in this situation, the school would sometimes forward the request packet to the district. Other times, the school returned the materials to us, and we sent them to the school district.
- ***The student transferred to another school.*** When the student had transferred to

---

sometimes due to such mismatches. In these cases, we attempted to learn the correct name and address of the school where the student was enrolled and make a new transcript request.

[52] For example, some school principals and registrars indicated that their school was often confused with another school having the same or a similar name and suggested that we direct our request to the other school. In this case, we would call the alternate school to find out if the student was ever enrolled there. If so, we made a correction to the database and sent a request to the newly identified school.

another school, a transcript was requested from the school to which the student had transferred. In some cases, the registrar or school secretary forwarded the request materials to the transfer school. in other cases, the request materials were sent back to us and we sent a new request to the transfer school.

- ***The school would not release any transcript without student's written consent.*** A few schools returned the transcript request materials with no transcripts, indicating that they required written consent from each student whose transcript we were requesting. A problem sheet was completed for these cases, and they were forwarded to the survey manager for follow-up. As a first step, the survey manager called the school to explain that, as an agent of the Department of Education, Mathematica Policy Research was authorized to collect student transcripts for the purposes of this study and that, according to the laws of *FERPA*, schools are permitted to release student transcripts to the Department of Education without the written consent of students participating in the study. It was also explained that students had given verbal consent over the telephone or written consent when they completed the mail survey, and that we did not request transcripts for any students who refused consent. Some schools agreed to send the requested transcripts upon hearing this explanation. Others reiterated that signed consent was required by school policy. in this case, we sent written consent forms to the students for them to sign and return to Mathematica so that we could obtain their student transcript for the impact study. A postage-paid return envelope was included with the consent form. A small number of students did sign and return the consent form, but most of the letters came back unopened because we no longer had a valid address for the student.
- ***The school would not release transcripts without advance payment.*** In these cases, we sent a check to cover the cost of each transcript, along with a list of the students whose transcripts we were requesting.
- ***The school would not release a transcript until the student paid an outstanding debt.*** In some cases we were eventually able to obtain these transcripts as students paid whatever bills they owed the school. When the debt remained unpaid, however, there was no way we could get the transcript. These cases were marked as unfilled requests.

4. Response Rates

From the samples used in the impact analysis described in Chapter III, 1,365 students reported having attended at least one postsecondary institution that could be matched to IPEDS. for each of these 1,365 students, we requested a transcript from each of the postsecondary institutions that he or she reported attending. In total, we requested 2,029 transcripts. We received 1,821 of the 2,029 transcripts requested (90 percent), and we obtained a complete transcript record—that is, transcripts for all postsecondary institutions attended—for 1,109 students (see Table A.2).

Table A.2. Postsecondary Transcript Data Collection for the Upward Bound Math-Science Impact Analysis

Postsecondary Students	UBMS Sample	Regular UB Sample	Full Sample
Complete Transcript Record	472	637	1,109
Incomplete Transcript Record	101	155	256
Total Postsecondary Students	573	792	1,365

APPENDIX B. PROGRAM IMPACTS

Table B.1. Impact of Upward Bound Math Science on High School Outcomes

	UBMS Participants	Comparison Group	Impact
Overall			
Total Credits	25.7	24.5	1.2 ***
Grade Point Average (GPA)	3.1	3.1	0.1 ***
Math Courses Taken			
Algebra (%)	97	97	0 !
Geometry (%)	86	86	0
Algebra II (%)	73	74	-1
Trigonometry (%)	35	30	5
Analysis / Precalculus (%)	36	31	5
Calculus (%)	26	24	2
Advanced Placement Calculus (%)	6	10	-3 *
GPA in Algebra or above	2.8	2.7	0.1 ***
Science Courses Taken			
Biology (%)	97	98	-1
Chemistry (%)	88	78	10 ***
Physics (%)	58	43	15 ***
Advanced Placement Biology, Chemistry, or Physics (%)	4	9	-5 !
GPA in Biology, Chemistry, and Physics	2.9	2.7	0.2 ***
High School Status (%)			
Graduated	99	96	2 ***
Dropped out	1	2	-2 !
General Educational Development (GED)	0	1	-1!

Source: MPR analysis file "UBMSImpact6.sas."

*/**/*** Impact estimate is statistically significant at the 0.10 /0.05 / 0.01 level.

! Statistical significance could not be assessed due to complete or semi-complete separation.

Note 1: All estimates are weighted to account for missing data.

Note 2: The comparison group estimates and impact estimates are regression-adjusted (see Chapter III, Section A.4).

Table B.2. Impact of Upward Bound Math Science on High School Outcomes by Prior Participation in Regular Upward Bound

	Participated in Regular Upward Bound			Did Not Participate in Regular Upward Bound			
	Partici-pants UBMS	Compar-ison Group	Impact	UBMS Participants	Comparison Group	Impact	
Overall							
Total Credits	24.6	24.5	0.1	26.0	24.5	1.5 ***	#
Grade Point Average (GPA)	3.1	3.0	0.0	3.2	3.1	0.1 ***	
Math Courses Taken							
Algebra (%)	97	98	-1!	97	97	1!	
Geometry (%)	88	88	0	86	85	1	
Algebra II (%)	66	78	-12 **	74	73	1	#
Trigonometry (%)	38	27	11 **	34	30	4	
Analysis / Precalculus (%)	34	29	5	37	32	5	
Calculus (%)	22	21	2	27	25	2	
Advanced Placement Calculus (%)	6	7	-2	7	10	-3 *	
GPA in Algebra or above	2.7	2.6	0.1	2.8	2.7	0.1 ***	
Science Courses Taken							
Biology (%)	98	97	1!	97	98	-1	
Chemistry (%)	91	78	13 ***	88	79	9 ***	
Physics (%)	53	41	12 **	59	43	16 ***	
Advanced Placement Biology, Physics (%)	2	5	-4 !	5	10	-5 **	
GPA in Biology, Chemistry, and Physics	2.8	2.7	0.1 **	2.9	2.7	0.2 ***	
High School Status (%)							
Graduated	100	99	1!	99	96	3 **	
Dropped out	0	1	-1!	1	3	-2 !	
General Educational Development (GED)	0	0	0!	1	1	-1!	

Source: MPR analysis file "UBMSImpact6.sas."

*/**/*** Impact estimate is statistically significant at the 0.10 /0.05 / 0.01 level.

Impact estimate is significantly different from the impact on students who participated in regular Upward Bound at the 0.10 level. ! Statistical significance could not be assessed due to complete or semi-complete separation.

Note 1: All estimates are weighted to account for missing data.

Note 2: The comparison group estimates and impact estimates are regression-adjusted (see Chapter III, Section A.4).

Table B.3. Impact of Upward Bound Math Science on High School Outcomes by Sex

	Male			Female		
	Partici-pants UBMS	Compar-ison Group	Impact	UBMS Participants	Comparison Group	Impact
Overall						
Total Credits	25.3	24.6	0.7	26.1	24.5	1.6 ***
Grade Point Average (GPA)	3.1	3.0	0.1 *	3.2	3.1	0.1 ***
Math Courses Taken						
Algebra (%)	97	98	0!	97	97	0!
Geometry (%)	84	87	-3	88	84	3
Algebra II (%)	77	73	4	70	74	-5
Trigonometry (%)	33	34	-1	36	28	8 *
Analysis / Precalculus (%)	42	35	7	32	28	4
Calculus (%)	25	22	3	27	25	2
Advanced Placement Calculus (%)	7	9	-2	6	10	-4
GPA in Algebra or above	2.7	2.6	0.1	2.8	2.7	0.1 ***
Science Courses Taken						
Biology (%)	98	99	-1!	97	98	-1
Chemistry (%)	87	77	11 ***	89	80	9 ***
Physics (%)	60	48	12 **	56	39	18 ***
Advanced Placement Biology, Chemistry, or Physics (%)	5	7	-2 !	4	10	-6 !
GPA in Biology, Chemistry, and Physics	2.9	2.6	0.3 ***	2.9	2.8	0.2 ***
High School Status (%)						
Graduated	98	98	0 !	99	96	4 !
Dropped out	1	1	0 !	0	2	-2 !
General Educational Development (GED)	1	1	0 !	0	2	-1!

Source: MPR analysis file "UBMSImpact6.sas."

*/**/*** Impact estimate is statistically significant at the 0.10 /0.05 / 0.01 level.

Impact estimate is significantly different from the impact on males at the 0.10 level.

! Statistical significance could not be assessed due to complete or semi-complete separation.

Note 1: All estimates are weighted to account for missing data.

Note 2: The comparison group estimates and impact estimates are regression-adjusted (see Chapter III, Section A.4).

Table B.4. Impact of Upward Bound Math Science on High School Outcomes by Race and Ethnicity

	African American			White			Hispanic		
UBMS Participants			Impact	UBMS Partici-pants	Compar-ison Group	Impact	UBMS Partici-pants	Compar-ison Group	Impact
Overall									
Total Credits	25.5	24.1	1.4 **	26.0	25.2	0.8	26.4	24.6	1.9 **
Grade Point Average (GPA)	2.9	2.9	0.1 *	3.3	3.3	0.0	3.2	3.0	0.2 ***#
Math Courses Taken									
Algebra (%)	96	97	-1!	100	99	1!	96	95	0 !
Geometry (%)	88	89	-1	91	87	5 **	81	79	2
Algebra II (%)	75	75	0	79	81	-2 !	67	74	-7
Trigonometry (%)	35	32	2	38	31	7	35	24	11 *
Analysis / Precalculus (%)	33	27	5	30	22	8	42	30	12
Calculus (%)	17	16	1	27	27	1	29	21	9 !
Advanced Placement Calculus (%)	4	4	0 !	4	12	-8 !	9	8	1!
GPA in Algebra or above	2.5	2.4	0.1	3.2	3.1	0.1	2.7	2.4	0.3 ***#
Science Courses Taken									
Biology (%)	97	99	-2	98	98	0 !	98	96	2 !
Chemistry (%)	90	83	7 **	87	77	10 !	87	70	17 ***#
Physics (%)	53	46	7	59	36	22 *** #	59	32	27 *** #
Advanced Placement Biology, Chemistry, or Physics (%)	4	7	-3 !	2	6	-4 !	6	12	-6 !
GPA in Biology, Chemistry, and Physics	2.7	2.5	0.2 ***	3.3	3.0	0.3 ***	2.9	2.7	0.2 **
High School Status (%)									
Graduated	100	97	3 !	99	97	2 !	98	96	3 !
Dropped out	0	2	-2 !	0	2	-2 !	2	4	-3 !
General Educational Development (GED)	0	0	0 !	1	1	0 !	0	2	-2 !

Source: MPR analysis file "UBMSImpact6.sas."

*/**/*** Impact estimate is statistically significant at the 0.10 /0.05 / 0.01 level.

Impact estimate is significantly different from the impact on African Americans at the 0.10 level.

! Statistical significance could not be assessed due to complete or semi-complete separation.

Note 1: All estimates are weighted to account for missing data.

Note 2: The comparison group estimates and impact estimates are regression-adjusted (see Chapter III, Section A.4).

Table B.5. Impact of Upward Bound Math Science on Postsecondary Outcomes

	UBMS Participants	Comparison Group	Impact
Postsecondary Enrollment (%)			
Any postsecondary institution	95	90	5 ***
Highest level of schooling attended			
Four-year college or university	82	71	11 ***
Two-year college	11	16	-5 **
Vocational institution	2	4	-2
College Selectivity (%)			
Most selective four-year college or university			
More selective	33	23	11 ***
Less selective	48	48	0
Postsecondary Credits Earned (mean)			
Two- and four-year colleges and universities	98.1	85.0	13.1***
Four-year colleges and universities	86.6	73.1	13.5 **
Two-year colleges and universities	11.5	11.9	-0.5
Postsecondary Completion (%)			
Any degree, certificate, or license	47	46	1
Highest degree, certificate, or license earned			
Bachelor's degree or higher	35	33	2
Associate's degree	7	7	-1
Certificate or license	5	5	-1

Source: MPR analysis file "UBMSImpact6.sas."

*/**/*** Impact estimate is statistically significant at the 0.10 / 0.05 / 0.01 level.

! Statistical significance could not be assessed due to complete or semi-complete separation.

Note 1: All estimates are weighted to account for missing data.

Note 2: The comparison group estimates and impact estimates are regression-adjusted (see Chapter III, Section A.4).

Table B.6. Impact of Upward Bound Math Science on Postsecondary Outcomes by Prior Participation in Regular Upward Bound

	Participated in Regular Upward Bound			Did Not Participate in Regular Upward Bound		
	UBMS Partici-pants	Compa-rison Group	Impact	UBMS Partici-pants	Compa-rison Group	Impact
Postsecondary Enrollment (%)						
Any postsecondary institution	99	91	8 ***	95	90	5 ***
Highest level of schooling attended						
Four-year college or university	81	71	10 *	82	70	12 ***
Two-year college	18	16	1	10	16	-6 **
Vocational institution	0	4	-4 !	3	4	-1
College Selectivity (%)						
Most selective four-year college or university						
More selective	35	24	11 **	33	22	11 ***
Less selective	46	47	-1	48	48	0
Postsecondary Credits Earned (mean)						
Two- and four-year colleges and universities	103.4	84.2	19.2 **	96.9	85.2	11.7 **
Four-year colleges and universities	90.1	69.2	20.9 **	85.9	74.3	11.5 *
Two-year colleges and universities	13.3	15.0	-1.7	11.1	10.9	0.2
Postsecondary Completion (%)						
Any degree, certificate, or license	42	43	-2	48	46	2
Highest degree, certificate, or license earned						
Bachelor's degree or higher	31	29	2	36	34	2
Associate's degree	7	5	2	7	8	-1
Certificate or license	4	8	-4	5	5	0

Source: MPR analysis file "UBMSImpact6.sas."

*/**/*** Impact estimate is statistically significant at the 0.10 / 0.05 / 0.01 level.

Impact estimate is significantly different from the impact on students who participated in regular Upward Bound at the 0.10 level.

! Statistical significance could not be assessed due to complete or semi-complete separation.

Note 1: All estimates are weighted to account for missing data.

Note 2: The comparison group estimates and impact estimates are regression-adjusted (see Chapter III, Section A.4).

Table B.7. Impact of Upward Bound Math Science on Postsecondary Outcomes by Sex

	Male			Female			
	UBMS Partici-pants	Comparison Group	Impact	UBMS Partici-pants	Comparison Group	Impact	
Postsecondary Enrollment (%)							
Any postsecondary institution	96	89	7 ***	95	90	5 ***	
Highest level of schooling attended							
Four-year college or university	82	76	6	82	68	14 ***	#
Two-year college	12	12	0	10	18	-8 ***	#
Vocational institution	2	2	0	2	4	-2	
College Selectivity (%)							
Most selective four-year college or university							
More selective	34	23	11 **	32	21	11 ***	
Less selective	48	52	-5	48	46	2	
Postsecondary Credits Earned (mean)							
Two- and four-year colleges and universities	96.6	85.7	10.9	99.1	85.2	13.9 ***	
Four-year colleges and universities	83.2	73.5	9.7	88.9	72.7	16.2 ***	
Two-year colleges and universities	13.4	12.2	1.2	10.1	12.5	-2.3	
Postsecondary Completion (%)							
Any degree, certificate, or license	40	43	-3	51	46	6	
Highest degree, certificate, or license earned							
Bachelor's degree or higher	27	33	-6	40	32	9 **	#
Associate's degree	8	4	4 *	6	10	-4 *	#
Certificate or license	5	5	0	5	5	0	

Source: MPR analysis file "UBMSImpact6.sas."

*/**/*** Impact estimate is statistically significant at the 0.10 / 0.05 / 0.01 level.

Impact estimate is significantly different from the impact on males at the 0.10 level.

Table B.8. Impact of Upward Bound Math Science on Postsecondary Outcomes by Race and Ethnicity

	African American			White			Hispanic		
	UBMS Partici-pants	Comparison Group	Impact	UBMS Participants	Comparison Group	Impact	UBMS Partici-pants	Comparison Group	Impact
Postsecondary Enrollment (%)									
Any postsecondary institution	97	93	5!	95	87	8 !	95	92	2 !
Highest level of schooling attended									
Four-year college or university	87	74	13 ***	79	67	12 !	80	72	8 !
Two-year college	9	15	-6	12	19	-7 !	14	17	-3 !
Vocational institution	1	3	-2 !	5	1	4 !	0	6	-6 !
College Selectivity (%)									
Most selective four-year college or university									
More selective	29	19	10 **	24	11	13 ***	38	36	3
Less selective	57	54	3	54	55	0	42	35	7
Postsecondary Credits Earned (mean)									
Two- and four-year colleges and universities	105.9	87.0	18.9 **	92.9	85.8	7.2	97.6	65.7	31.9 ***
Four-year colleges and universities	99.4	79.2	20.2 **	81.6	73.3	8.4	81.2	52.9	28.3 ***
Two-year colleges and universities	6.5	7.8	-1.3	11.3	12.5	-1.2	16.4	12.8	3.6
Postsecondary Completion (%)									
Any degree, certificate, or license	49	46	3	46	35	10 **	49	49	-1
Highest degree, certificate, or license earned									
Bachelor's degree or higher	38	36	1	36	19	17 *** #	33	30	3
Associate's degree	5	4	2 !	8	10	-3	8	13	-5
Certificate or license	6	6	-1!	2	6	-4 !	8	5	3

Source: MPR analysis file "UBMSImpact6.sas."

*/**/*** Impact estimate is statistically significant at the 0.10 / 0.05 / 0.01 level.

Impact estimate is significantly different from the impact on African Americans at the 0.10 level. ! Statistical significance could not be assessed due to complete or semi-complete separation; Note 1: All estimates are weighted to account for missing data; Note 2: The comparison group estimates and impact estimates are regression-adjusted (see Chapter III, Section A.4).

Table B.9. Impact of Upward Bound Math Science on Postsecondary Field of Study

Field of Study	UBMS Participants	Comparison Group	Impact	
Field of Study at Most Recent PS Institution (%)				
All postsecondary institutions				
Math and science fields	33	23	10	***
Social science fields	11	7	4	*
Other fields	42	51	-9	* **
Four-year colleges and universities				
Math and science fields	28	18	10	***
Social science fields	10	7	4	*
Other fields	36	39	-3	
Earned Degree or Certificate in Field (%)				
All postsecondary institutions				
Math and science fields	15	12	3	
Social science fields	7	4	3	
Other fields	21	28	-7	**
Four-year colleges and universities				
Math and science fields	12	6	6	***
Social science fields	6	4	3	
Other fields	14	20	-6	**

Source: MPR analysis file "UBMSImpact6.sas."

*/**/*** Impact estimate is statistically significant at the 0.10 /0.05 / 0.01 level.

! Statistical significance could not be assessed due to complete or semi-complete separation.

Note 1: All estimates are weighted to account for missing data.

Note 2: The comparison group estimates and impact estimates are regression-adjusted (see Chapter III, Section A.56).

Table B.10. Impact of Upward Bound Math Science on Postsecondary Field of Study by Prior Participation in Regular Upward Bound

	Participated in Regular Upward Bound			Did Not Participate in Regular Upward Bound		
Field of Study	UBMS Participants	Comparison Group	Impact	UBMS Participants	Comparison Group	Impact
Field of Study at Most Recent PS Institution (%)						
All postsecondary institutions						
Math and science fields	35	26	10 *	32	22	10 **
Social science fields	13	7	6 *	11	8	3
Other fields	38	47	-9	43	52	-9 **
Four-year colleges and universities						
Math and science fields	29	23	6	28	17	12 ***
Social science fields	12	6	6	10	7	3
Other fields	31	35	-3	37	40	-4

Table B.10. Continued

	Participated in Regular Upward Bound			Did Not Participate in Regular Upward Bound		
Field of Study	UBMS Participants	Comparison Group	Impact	UBMS Participants	Comparison Group	Impact
Earned Degree or Certificate in Field (%)						
All postsecondary institutions						
Math and science fields	13	12	1	15	11	4
Social science fields	6	3	3	7	4	3
Other fields	20	24	-4	22	29	-7 **
Four-year colleges and universities						
Math and science fields	8	10	-2 !	13	5	8 ***
Social science fields	4	3	1	7	4	3
Other fields	15	13	2	14	22	-8 **

Source: UBMSImpact6.sas

*/**/*** Impact estimate is statistically significant at the 0.10 /0.05 / 0.01 level.

Impact estimate is significantly different from the impact on students who participated in regular Upward Bound at the 0.10 level. ! Statistical significance could not be assessed due to complete or semi-complete separation.

Note 1: All estimates are weighted to account for missing data.

Note 2: The comparison group estimates and impact estimates are regression-adjusted (see Chapter III, Section A.4).

Table B.11. Impact of Upward Bound Math Science on Postsecondary Field of Study by Sex

	Male			Female			
Field of Study	UBMS Participants	Compari-son Group	Impact	UBMS Participants	Compari-son Group	Impact	
Field of Study at Most Recent PS Institution (%)							
All postsecondary institutions							
Math and science fields	44	28	16 ***	24	19	5	
Social science fields	9	5	4	12	9	3	
Other fields	31	46	-15***	50	53	-3	#
Four-year colleges and universities							
Math and science fields	38	25	13 **	22	13	9 **	
Social science fields	8	4	5 *	12	9	3	
Other fields	27	38	-12 **	42	39	4	#
Earned Degree or Certificate in Field (%)							
All postsecondary institutions							
Math and science fields	16	10	6 *	14	11	3	
Social science fields	5	3	2 !	8	5	3 *	
Other fields	15	25	-10 **	26	28	-2	
Four-year colleges and universities							
Math and science fields	11	6	5 *	13	5	8 ***	
Social science fields	4	3	1!	8	4	4 **	
Other fields	9	19	-11 ***	18	20	-2	#

Source: MPR analysis file "UBMSImpact6.sas."

*/**/*** Impact estimate is statistically significant at the 0.10 /0.05 / 0.01 level.

Impact estimate is significantly different from the impact on males at the 0.10 level.

Table B.12. Impact of Upward Bound Math Science on Postsecondary Field of Study by Race and Ethnicity

Field of Study	African American UBMS Participants	African American Compari-son Group	African American Impact	White UBMS Partici-pants	White Compari-son Group	White Im-pact	Hispanic UBMS Partici-pants	Hispanic Compari-son Group	Hispanic Impact
Field of Study at Most Recent PS Institution (%)									
All postsecondary institutions									
Math and science fields	32	25	7	34	26	8	28	14	15 **
Social science fields	13	6	6 **	9	4	5	12	8	4 !
Other fields	42	59	-17 ***	42	50	-8	48	55	-7
Four-year colleges and universities									
Math and science fields	30	20	10 *	27	20	7	26	8	18 **
Social science fields	13	6	7 **	8	4	4 !	8	7	1!
Other fields	38	44	-7	35	38	-4	40	45	-5
Earned Degree or Certificate in Field (%)									
All postsecondary institutions									
Math and science fields	14	10	4 !	16	9	6	15	9	7
Social science fields	8	4	4 *	5	2	4 !	5	5	0 !
Other fields	21	32	-11 **	21	20	1 #	26	30	-4
Four-year colleges and universities									
Math and science fields	13	6	7 !	13	5	8 !	11	2	9 !
Social science fields	7	3	4 *	5	2	4 !	5	5	0 !
Other fields	15	23	-8 *	13	9	4 #	15	19	-4

Source: MPR analysis file "UBMSImpact6.sas."

*/**/*** Impact estimate is statistically significant at the 0.10 /0.05 / 0.01 level.

Impact estimate is significantly different from the impact on African Americans at the 0.10 level. ! Statistical significance could not be assessed due to complete or semi-complete separation.

Note 1: All estimates are weighted to account for missing data.

Note 2: The comparison group estimates and impact estimates are regression-adjusted (see Chapter III, Section A.4).

Appendix C. Sensitivity of Impacts to an Alternative Measure of Postsecondary Attendance

A. Verification of Students' Postsecondary Attendance

In Chapter III, we presented estimates of the effects of UBMS on postsecondary attendance based on self-reported attendance—specifically, the schools that sample members reported attending after high school. However, in some instances, sample members may not have actually attended the schools they reported. For example, some may have reported schools that they planned to attend but never attended, and others may have reported schools at which they participated in a noncredit program but were not enrolled as students. In this appendix, we conduct an analysis to determine if the impact estimates presented in Chapter III are sensitive to an alternative measure of postsecondary attendance, one that requires verification by the postsecondary institution that students reported attending.

Myers et al. (2004) present the results of this analysis in the text of the report along with the analysis based on self-reported data, which effectively provided two impact estimates for each outcome. When one of the two impact estimates was statistically insignificant, the results were characterized as inconclusive because at that time, we were unable to persuade ourselves that one method was more reliable than the other. However, we have since conducted an investigation of the verification process, and the results of this analysis indicate that on average, the self-reported measures of postsecondary attendance are more accurate than the measures requiring verification by the institution. While this may seem counterintuitive, in constructing postsecondary attendance measures that required verification, we assumed that students did not attend the schools they reported unless attendance was verified by the school, typically through the provision of a transcript. However, the transcript collection process was not originally designed to verify attendance.[53] And in most instances when a school was unable or unwilling to provide a transcript, a careful examination of the data suggests that the student probably did attend the school—or at the very least, the school gave us no reason to doubt the student had attended when explaining why they would not provide a transcript.

The verification process worked as follows. In the third and fourth follow-up surveys, students reported all postsecondary institutions that they had attended since high school. We then attempted to match all reported schools to the 1997-98 Integrated Postsecondary Education Data System (IPEDS) maintained by the National Center for Education Statistics (NCES) to determine whether they met NCES's definition of a postsecondary institution and to obtain school contact information. Transcript requests were then sent to all schools a student reported attending that we were able to match to IPEDS and schools that did not initially respond were followed up. If a school provided a transcript that we requested for a student in the sample, then the student clearly attended that school. Furthermore, some of the reasons given by school staff for not providing transcripts can be treated as verification of attendance, for example, a college indicating that it could not provide a transcript for a

[53] For the final report of this evaluation, which will use postsecondary transcript data collected in 2004-05, the transcript data collection process has been modified to specifically verify attendance if possible.

student because he or she owed money to the school. In many instances, however, the reason given for not providing transcripts does not clearly indicate whether the student attended the school. For example, some schools required written consent from the students themselves even though the law does not require it, and we typically obtained only oral consent (see Appendix B); in these cases, lack of verification casts no doubt on the student's self-reported attendance.

The information obtained while collecting transcripts is therefore useful in verifying attendance in some but not all cases. In the vast majority of cases considered (approximately 80 percent) we were able to verify the student's attendance. However, for approximately 220 students, the fact that we were unable to verify attendance for at least one of their reported postsecondary institutions changed at least one outcome of interest.[54] In investigating these cases, we found little evidence that contradicted a student's report of their postsecondary schools attended. One concern was that some Upward Bound participants would report the school at which they attended an Upward Bound summer bridge program but did not actually enroll; we found no evidence that this happened.[55] Similarly, there was little evidence that many students reported attending institutions that are not classified by IPEDS as postsecondary institutions, and, in fact, non-IPEDS institutions such as Job Corps programs were excluded through the IPEDS match. The lack of verification more often appeared to be due to schools that would not release transcripts without written consent, ambiguities in the exact campus attended by the student within a large state system, or insufficient information about the school reported by the student to know which school to contact for transcript information. More often than not, lack of verification seemed to mainly reflect limitations of the verification process rather than inaccuracies in the information provided by students.

B. Differences in Impacts

This section presents estimates of the impacts of Upward Bound that use information only on schools that we were able to verify the student attended. These estimates make the strong assumption that lack of verification of attendance implies that the student did not attend the school in question or any school like it.

Because most attendance reported by students was verified through transcript receipt, most of the impacts presented here are very similar to those estimated using information on all schools that students reported attending. There are a few differences between the estimates presented here and the estimates presented in Chapter III, but there is probably only one worth noting: the finding that UBMS has a positive effect on the likelihood of attending some type of postsecondary institution is sensitive to the verification process. If we require

[54] There may be cases in which we were not able to verify a student's attendance at all schools but for which that did not affect the outcome variables of interest. For example, if we verified a student's attendance at a four-year college or university but could not verify his or her attendance at a vocational school they reported attending, he or she will still be classified as having their highest level of schooling as a four-year college or university.

[55] Students participate in Upward Bound summer bridge programs the summer before enrolling in college. Because the survey question simply asks students to list all postsecondary

verification of postsecondary attendance, the estimated effect becomes statistically insignificant (compare Table C.1 to Table B.5). For the full set of estimates from the sensitivity analysis, see Tables C.1 - C.4.

Table C.1. Impact of Upward Bound Math Science on Postsecondary Outcomes, Excludes Unverified Enrollment and Completion

	UBMS Participants	Comparison Group	Impact
Postsecondary Enrollment (%)			
Any postsecondary institution	89	87	2
Highest level of schooling attended			
Four-year college or university	77	68	9***
Two-year college	11	16	-5 **
Vocational institution	1	3	-2 *
College Selectivity (%)			
Most selective four-year college or university			
More selective	30	21	9 ***
Less selective	46	47	-1
Postsecondary Credits Earned (mean)			
Two- and four-year colleges and universities	84.8	75.2	9.6 **
Four-year colleges and universities	75.3	64.8	10.5 **
Two-year colleges and universities	9.6	10.4	-0.8
Postsecondary Completion (%)			
Any degree, certificate, or license	43	43	0
Highest degree, certificate, or license earned			
Bachelor's degree or higher	33	32	1
Associate's degree	6	7	-1
Certificate or license	4	4	1

Source: MPR analysis file "UBMSImpact6.sas."

*/**/*** Impact estimate is statistically significant at the 0.10 / 0.05 / 0.01 level.

! Statistical significance could not be assessed due to complete or semi-complete separation.

Note 1: All estimates are weighted to account for missing data.

Note 2: The comparison group estimates and impact estimates are regression-adjusted (see Chapter III, Section A.4).

---

institutions attended since high school graduation, there was concern that some students may list the school at which they participated in a summer bridge program.

Table C.2. Impact of Upward Bound Math Science on Postsecondary Outcomes by Prior Participation in Regular Upward Bound, Excludes Unverified Enrollment and Completion

	Participated in Regular Upward Bound			Did Not Participate in Regular Upward Bound		
	UBMS Participants	Comparison Group	Impact	UBMS Participants	Comparison Group	Impact
Postsecondary Enrollment (%)						
Any postsecondary institution	91	86	4	89	87	2
Highest level of schooling attended						
Four-year college or university	75	67	8	77	68	9 ***
Two-year college	16	17	0	10	16	-6 **
Vocational institution	0	2	-2 !	1	3	-2
College Selectivity (%)						
Most selective four-year college or university						
More selective	31	23	8	30	20	10 ***
Less selective	44	43	0	46	47	-1
Postsecondary Credits Earned (mean)						
Two- and four-year colleges and universities	83.5	65.7	17.8 ***	85.1	77.1	8.1
Four-year colleges and universities	72.8	53.9	18.9 **	75.8	67.0	8.8
Two-year colleges and universities	10.7	11.7	-1.1	9.3	10.1	-0.7

Table C.2. Continued

	Participated in Regular Upward Bound			Did Not Participate in Regular Upward Bound		
	UBMS Participants	Comparison Group	Impact	UBMS Participants	Comparison Group	Impact
Postsecondary Completion (%)						
Any degree, certificate, or license	36	41	-6	45	43	2
Highest degree, certificate, or license earned						
Bachelor's degree or higher	25	28	-2	35	33	2
Associate's degree	6	5	1	6	7	-1
Certificate or license	4	7	-2	4	3	1

Source: MPR analysis file "UBMSImpact6.sas."

*/**/*** Impact estimate is statistically significant at the 0.10 / 0.05 / 0.01 level.

Impact estimate is significantly different from the impact on students who participated in regular Upward Bound at the 0.10 level. ! Statistical significance could not be assessed due to complete or semi-complete separation.

Note 1: All estimates are weighted to account for missing data.

Note 2: The comparison group estimates and impact estimates are regression-adjusted (see Chapter III, Section A.4).

Table C.3. Impact of Upward Bound Math Science on Postsecondary Outcomes by Sex, Excludes Unverified Enrollment and Completion

	Male			Female			
	UBMS Participants	Comparison Group	Impact	UBMS Participants	Comparison Group	Impact	
Postsecondary Enrollment (%)							
Any postsecondary institution	88	88	1	89	85	4 *	
Highest level of schooling attended							
Four-year college or university	74	74	0	78	64	15 ***	#
Two-year college	13	13	0	10	18	-8 ***	#
Vocational institution	1	1	0 !	1	4	-3 *	
College Selectivity (%)							
Most selective four-year college or university							
More selective	30	21	9 **	30	19	10 ***	
Less selective	43	53	-9 *	47	44	3	#
Postsecondary Credits Earned (mean)							
Two- and four-year colleges and universities	80.2	76.3	3.9	88.1	73.9	14.2 ***	
Four-year colleges and universities	69.7	67.6	2.0	79.2	62.0	17.3 ***	#
Two-year colleges and universities	10.6	8.7	1.9	8.9	11.9	-3.1	

Table C.3. Continued

	Male			Female			
	UBMS Participants	Comparison Group	Impact	UBMS Participants	Comparison Group	Impact	
Postsecondary Completion (%)							
Any degree, certificate, or license	37	42	-5	48	43	6	#
Highest degree, certificate, or license earned							
Bachelor's degree or higher	26	33	-7	38	31	7 *	#
Associate's degree	7	3	4 *	6	10	-4 *	#
Certificate or license	4	4	0	4	3	1	

Source: MPR analysis file "UBMSImpact6.sas."

*/**/*** Impact estimate is statistically significant at the 0.10 / 0.05 / 0.01 level.

Impact estimate is significantly different from the impact on males at the 0.10 level.

! Statistical significance could not be assessed due to complete or semi-complete separation.

Note 1: All estimates are weighted to account for missing data.

Note 2: The comparison group estimates and impact estimates are regression-adjusted (see Chapter III, Section A.4).

Table C.4. Impact of Upward Bound Math Science on Postsecondary Outcomes by Race and Ethnicity, Excludes Unverified Enrollment and Completion

	African American			White			Hispanic		
	UBMS Participants	Comparison Group	Impact	UBMS Participants	Comparison Group	Impact	UBMS Participants	Comparison Group	Impact
Postsecondary Enrollment (%)									
Any postsecondary institution	91	89	1!	89	86	4 !	87	83	4 !
Highest level of schooling attended									
Four-year college or university	82	72	11 **	74	67	7 !	72	62	10 !
Two-year college	8	16	-8 **	14	19	-5 !	16	19	-3 !
Vocational institution	1	3	-2 !	2	1	1!	0	5	-5 !
College Selectivity (%)									
Most selective four-year college or university									
More selective	26	17	9 **	23	11	12 ***	34	25	9
Less selective	55	54	1	50	54	-5	38	34	4
Postsecondary Credits Earned (mean)									
Two- and four-year colleges and universities	88.5	73.8	14.8 **	86.2	77.2	9.0	82.4	52.6	29.8 ***
Four-year colleges and universities	83.3	66.7	16.7 **	75.9	65.7	10.2	67.9	41.3	26.6 ***
Two-year colleges and universities	5.2	7.1	-1.9	10.3	11.4	-1.2	14.5	11.4	3.2

Table C.4. Continued

	African American			White			Hispanic		
	UBMS Participants	Comparison Group	Impact	UBMS Participants	Comparison Group	Impact	UBMS Participants	Comparison Group	Impact
Postsecondary Completion (%)									
Any degree, certificate, or license	44	41	3	43	35	9 *	46	43	3
Highest degree, certificate, or license earned									
Bachelor's degree or higher	35	36	-1	35	19	16 *** #	32	25	6
Associate's degree	4	4	1!	8	10	-2	7	13	-6
Certificate or license	5	2	3	1	6	-5 !	8	3	4

Source: MPR analysis file "UBMSImpact6.sas."

*/**/*** Impact estimate is statistically significant at the 0.10 / 0.05 / 0.01 level.

Impact estimate is significantly different from the impact on African Americans at the 0.10 level. ! Statistical significance could not be assessed due to complete or semi-complete separation.

Note 1: All estimates are weighted to account for missing data.

Note 2: The comparison group estimates and impact estimates are regression-adjusted (see Chapter III, Section A.4).

APPENDIX D. SAMPLE SIZES AND STANDARD ERRORS

Table D.1. Sample Sizes and Standard Errors for Reported Impact Estimates: Table B. 1

	Sample Size	Standard Error
Overall		
Total Credits	1,677	0.40
Grade Point Average (GPA)	1,677	0.02
Math Courses Taken		
Algebra (%)	1,677	0.70
Geometry (%)	1,677	2.35
Algebra II (%)	1,677	3.80
Trigonometry (%)	1,677	4.16
Analysis / Precalculus (%)	1,677	4.53
Calculus (%)	1,677	2.77
Advanced Placement Calculus (%)	1,677	1.43
GPA in Algebra or above	1,677	0.05
Science Courses Taken		
Biology (%)	1,677	0.91
Chemistry (%)	1,677	1.61
Physics (%)	1,677	3.07
Advanced Placement Biology, Chemistry, or Physics (%)	1,677	1.38
GPA in Biology, Chemistry, and Physics	1,677	0.04
High School Status (%)		
Graduated	1,677	0.51
Dropped out	1,677	0.37
General Educational Development (GED)	1,677	0.38

Source: MPR analysis file "UBMSImpact6.sas."

Note: Standard errors account for project clustering and were estimated using Taylor series linearization methods.

Table D.2. Sample Sizes and Standard Errors for Reported Impact Estimates: Table B.2

	Participated in Regular Upward		Did Not Participate in Regular	
	Sample Size	Standard Error	Sample Size	Standard Error
Overall				
Total Credits	640	0.52	1,037	0.43
Grade Point Average (GPA)	640	0.03	1,037	0.03
Math Courses Taken				
Algebra (%)	640	1.10	1,037	0.63
Geometry (%)	640	2.84	1,037	2.57
Algebra II (%)	640	5.85	1,037	3.66
Trigonometry (%)	640	6.08	1,037	4.59
Analysis / Precalculus (%)	640	5.55	1,037	4.91
Calculus (%)	640	4.21	1,037	3.17
Advanced Placement Calculus (%)	640	2.55	1,037	1.69
GPA in Algebra or above	640	0.08	1,037	0.05
Science Courses Taken				
Biology (%)	640	1.05	1,037	1.05
Chemistry (%)	640	2.44	1,037	1.88
Physics (%)	640	5.27	1,037	3.43
Advanced Placement Biology, Chemistry, or Physics (%)	640	1.44	1,037	1.66
GPA in Biology, Chemistry, and Physics	640	0.06	1,037	0.05

Table D.2. Continued

	Participated in Regular Upward		Did Not Participate in Regular	
	Sample Size	Standard Error	Sample Size	Standard Error
High School Status (%)				
Graduated	640	0.00	1,037	0.64
Dropped out	640	0.00	1,037	0.47
General Educational Development (GED)	640	0.00	1,037	0.49

Source: MPR analysis file "UBMSImpact6.sas."

Note: Standard errors account for project clustering and were estimated using Taylor series linearization methods.

Table D.3. Sample Sizes and Standard Errors for Reported Impact Estimates: Table B.3

	Male		Female	
	Sample Size	Standard Error	Sample Size	Standard Error
Overall				
Total Credits	578	0.47	1,098	0.47
Grade Point Average (GPA)	578	0.04	1,098	0.03
Math Courses Taken				
Algebra (%)	578	1.05	1,098	0.77
Geometry (%)	578	2.91	1,098	2.49
Algebra II (%)	578	4.40	1,098	4.43
Trigonometry (%)	578	5.84	1,098	4.64
Analysis / Precalculus (%)	578	4.87	1,098	5.10
Calculus (%)	578	4.47	1,098	3.29
Advanced Placement Calculus (%)	578	2.62	1,098	1.90
GPA in Algebra or above	578	0.08	1,098	0.05
Science Courses Taken				
Biology (%)	578	1.04	1,098	1.14
Chemistry (%)	578	2.62	1,098	1.88
Physics (%)	578	5.19	1,098	4.10
Advanced Placement Biology, Chemistry, or Physics (%)	578	2.23	1,098	1.68
GPA in Biology, Chemistry, and Physics	578	0.07	1,098	0.05

Table D.3. Continued

	Male		Female	
	Sample Size	Standard Error	Sample Size	Standard Error
High School Status (%)				
Graduated	578	0.84	1,098	0.66
Dropped out	578	0.88	1,098	0.32
General Educational Development (GED)	578	0.09	1,098	0.59

Source: MPR analysis file "UBMSImpact6.sas."

Note: Standard errors account for project clustering and were estimated using Taylor series linearization methods.

Table D.4. Sample Sizes and Standard Errors for Reported Impact Estimates: Table B.4

	African American		White		Hispanic	
	Sample Size	Standard Error	Sample Size	Standard Error	Sample Size	Standard Error
Overall						
Total Credits	633	0.55	522	0.82	306	0.92
Grade Point Average (GPA)	633	0.04	522	0.03	306	0.05
Math Courses Taken						
Algebra (%)	633	1.19	522	0.00	306	2.46
Geometry (%)	633	2.82	522	2.07	306	5.01
Algebra II (%)	633	5.78	522	5.67	306	5.66
Trigonometry (%)	633	5.48	522	7.27	306	7.21
Analysis / Precalculus (%)	633	6.47	522	6.71	306	7.97
Calculus (%)	633	3.87	522	4.23	306	7.37
Advanced Placement Calculus (%)	633	2.46	522	1.78	306	4.31
GPA in Algebra or above	633	0.08	522	0.06	306	0.09
Science Courses Taken						
Biology (%)	633	1.69	522	1.33	306	1.19
Chemistry (%)	633	2.67	522	3.26	306	4.24
Physics (%)	633	5.79	522	6.16	306	7.06
Advanced Placement Biology, Chemistry, or Physics (%)	633	2.43	522	2.22	306	3.35
GPA in Biology, Chemistry, and Physics	633	0.06	522	0.06	306	0.12
High School Status (%)						
Graduated	633	0.00	522	1.09	306	1.46
Dropped out	633	0.00	522	0.00	306	1.65
General Educational Development (GED)	633	0.00	522	0.84	306	0.00

Source: MPR analysis file "UBMSImpact6.sas."

Note: Standard errors account for project clustering and were estimated using Taylor series linearization methods.

Table D.5. Sample Sizes and Standard Errors for Reported Impact Estimates: Table B.5

	Sample Size	Standard Error
Postsecondary Enrollment (%)		
Any postsecondary institution	1,438	0.99
Highest level of schooling attended		
Four-year college or university	1,438	2.28
Two-year college	1,438	1.99
Vocational institution	1,438	0.96
College Selectivity (%)		
Most selective four-year college or university		
More selective	1,438	3.44
Less selective	1,438	4.05
Postsecondary Credits Earned (mean)		
Two- and four-year colleges and universities	1,158	5.0
Four-year colleges and universities	1,158	5.7
Two-year colleges and universities	1,158	2.3
Postsecondary Completion (%)		
Any degree, certificate, or license	1,438	3.54
Highest degree, certificate, or license earned		
Bachelor's degree or higher	1,438	3.70
Associate's degree	1,438	1.70
Certificate or license	1,438	1.52

Source: MPR analysis file "UBMSImpact6.sas."

Note: Standard errors account for project clustering and were estimated using Taylor series linearization methods.

Table D.6. Sample Sizes and Standard Errors for Reported Impact Estimates: Table B.6

	Participated in Regular Upward		Did Not Participate in Regular	
	Sample Size	Standard Error	Sample Size	Standard Error
Postsecondary Enrollment (%)				
Any postsecondary institution	552	1.15	886	1.18
Highest level of schooling attended				
Four-year college or university	552	4.54	886	2.47
Two-year college	552	4.89	886	2.01
Vocational institution	552	0.00	886	1.30
College Selectivity (%)				
Most selective four-year college or university				
More selective	552	5.56	886	3.79
Less selective	552	5.88	886	4.49
Postsecondary Credits Earned (mean)				
Two- and four-year colleges and universities	432	7.40	726	5.38
Four-year colleges and universities	432	8.73	726	6.24
Two-year colleges and universities	432	4.71	726	2.28

Table D.6. Continued

	Participated in Regular Upward		Did Not Participate in Regular	
	Sample Size	Standard Error	Sample Size	Standard Error
Postsecondary Completion (%)				
Any degree, certificate, or license	552	5.88	886	3.86
Highest degree, certificate, or license earned				
Bachelor's degree or higher	552	5.19	886	4.14
Associate's degree	552	2.77	886	1.92
Certificate or license	552	2.(	886	1.

Source: MPR analysis file "UBMSImpact6.sas."

Note: Standard errors account for project clustering and were estimated using Taylor series linearization methods.

Table D.7. Sample Sizes and Standard Errors for Reported Impact Estimates: Table B.7

	Male		Female	
	Sample Size	Standard Error	Sample Size	Standard Error
Postsecondary Enrollment (%)				
Any postsecondary institution	480	1.62	957	1.31
Highest level of schooling attended				
Four-year college or university	480	3.66	957	2.76
Two-year college	480	3.80	957	2.29
Vocational institution	480	1.11	957	1.32
College Selectivity (%)				
Most selective four-year college or university				
More selective	480	5.25	957	3.65
Less selective	480	5.76	957	4.48
Postsecondary Credits Earned (mean)				
Two- and four-year colleges and universities	377	7.80	780	4.85
Four-year colleges and universities	377	8.97	780	5.47
Two-year colleges and universities	377	4.73	780	2.43
Postsecondary Completion (%)				
Any degree, certificate, or license	480	5.17	957	4.02
Highest degree, certificate, or license earned				
Bachelor's degree or higher	480	4.70	957	3.97
Associate's degree	480	3.35	957	1.74
Certificate or license	480	2.43	957	1.74

Source: MPR analysis file "UBMSImpact6.sas."

Note: Standard errors account for project clustering and were estimated using Taylor series linearization methods.

Table D.8. Sample Sizes and Standard Errors for Reported Impact Estimates: Table B.8

	African American		White		Hispanic	
	Sample Size	Standard Error	Sample Size	Standard Error	Sample Size	Standard Error
Postsecondary Enrollment (%)						
Any postsecondary institution	523	1.46	468	1.99	261	3.08
Highest level of schooling attended						
Four-year college or university	523	3.28	468	4.52	261	3.87
Two-year college	523	3.37	468	3.98	261	3.60
Vocational institution	523	1.00	468	2.58	261	0.01
College Selectivity (%)						
Most selective four-year college or university						
More selective	523	5.41	468	5.92	261	7.61
Less selective	523	6.33	468	6.07	261	7.95
Postsecondary Credits Earned (mean)						
Two- and four-year colleges and universities	405	7.24	394	8.10	211	9.16
Four-year colleges and universities	405	8.08	394	9.81	211	9.57
Two-year colleges and universities	405	2.63	394	4.33	211	5.37
Postsecondary Completion (%)						
Any degree, certificate, or license	523	6.30	468	5.01	261	6.62
Highest degree, certificate, or license earned						
Bachelor's degree or higher	523	6.09	468	4.96	261	6.59
Associate's degree	523	3.29	468	3.46	261	4.31
Certificate or license	523	2.50	468	1.23	261	3.66

Source: MPR analysis file "UBMSImpact6.sas."

Note: Standard errors account for project clustering and were estimated using Taylor series linearization methods.

Table D.9. Sample Sizes and Standard Errors for Reported Impact Estimates: Table B.9

	Sample Size	Standard Error
Field of Study at Most Recent PS Institution		
All PS institutions		
Math and science fields	1,438	3.73
Social science fields	1,438	2.37
Other fields	1,438	3.33
Four-year colleges and universities		
Math and science fields	1,438	3.53
Social science fields	1,438	2.37
Other fields	1,438	3.23
Earned Degree or Certificate in Field		
All PS institutions		
Math and science fields	1,438	2.84
Social science fields	1,438	2.13
Other fields	1,438	2.69
Four-year colleges and universities		
Math and science fields	1,438	2.78
Social science fields	1,438	2.17
Other fields	1,438	2.56

Source: MPR analysis file "UBMSImpact6.sas."

Note: Standard errors account for project clustering and were estimated using Taylor series linearization methods.

Table D.10. Sample Sizes and Standard Errors for Reported Impact Estimates: Table B.10

	Participated in Regular Upward		Did Not Participate in Regular	
	Sample Size	Standard Error	Sample Size	Standard Error
Field of Study at Most Recent PS Institution All PS institutions				
Math and science fields	552	5.98	886	4.48
Social science fields	552	4.70	886	2.64
Other fields	552	6.10	886	4.01
Four-year colleges and universities				
Math and science fields	552	5.76	886	4.21
Social science fields	552	4.55	886	2.63
Other fields	552	5.72	886	3.74
Earned Degree or Certificate in Field				
All PS institutions				
Math and science fields	552	3.99	886	3.34
Social science fields	552	2.90	886	2.49
Other fields	552	4.59	886	3.15
Four-year colleges and universities				
Math and science fields	552	2.89	886	3.59
Social science fields	552	2.94	886	2.67
Other fields	552	3.75	886	2.79

Source: MPR analysis file "UBMSImpact6.sas."

Note: Standard errors account for project clustering and were estimated using Taylor series linearization methods.

Table D.11. Sample Sizes and Standard Errors for Reported Impact Estimates: Table B.11

	Male		Female	
	Sample Size	Standard Error	Sample Size	Standard Error
Field of Study at Most Recent PS Institution All PS institutions				
Math and science fields	480	6.16	957	4.14
Social science fields	480	4.21	957	2.57
Other fields	480	4.81	957	4.09
Four-year colleges and universities				
Math and science fields	480	5.94	957	4.11
Social science fields	480	3.72	957	2.60
Other fields	480	4.55	957	3.76
Earned Degree or Certificate in Field				
All PS institutions				
Math and science fields	480	3.87	957	3.50
Social science fields	480	2.42	957	2.49
Other fields	480	3.69	957	3.44
Four-year colleges and universities				
Math and science fields	480	3.66	957	3.86
Social science fields	480	2.18	957	2.66
Other fields	480	2.84	957	3.01

Source: MPR analysis file "UBMSImpact6.sas."

Note: Standard errors account for project clustering and were estimated using Taylor series linearization methods.

Table D.12. Sample Sizes and Standard Errors for Reported Impact Estimates: Table B. 12

	African American		White		Hispanic	
	Sample Size	Standard Error	Sample Size	Standard Error	Sample Size	Standard Error
Field of Study at Most Recent PS Institution All PS institutions						
Math and science fields	523	5.84	468	6.00	261	7.31
Social science fields	523	3.42	468	3.23	261	4.95
Other fields	523	5.26	468	5.09	261	5.92
Four-year colleges and universities						
Math and science fields	523	6.25	468	4.94	261	11.09
Social science fields	523	3.63	468	3.18	261	3.70
Other fields	523	4.68	468	4.48	261	5.58
Earned Degree or Certificate in Field						
All PS institutions						
Math and science fields	523	4.64	468	5.17	261	7.69
Social science fields	523	3.11	468	3.08	261	3.26
Other fields	523	4.70	468	3.67	261	5.46
Four-year colleges and universities						
Math and science fields	523	5.85	468	4.82	261	6.85
Social science fields	523	3.06	468	3.08	261	3.65
Other fields	523	4.10	468	3.66	261	4.73

Source: MPR analysis file "UBMSImpact6.sas."

Note: Standard errors account for project clustering and were estimated using Taylor series linearization methods.

Table D.13. Sample Sizes and Standard Errors for Reported Impact Estimates: Table C. 1

	Sample Size	Standard Error
Postsecondary Enrollment (%)		
Any postsecondary institution	1,438	1.75
Highest level of schooling attended		
Four-year college or university	1,438	2.54
Two-year college	1,438	1.97
Vocational institution	1,438	0.70
College Selectivity (%)		
Most selective four-year college or university		
More selective	1,438	3.34
Less selective	1,438	3.78
Postsecondary Credits Earned (mean)		
Two- and four-year colleges and universities	1,438	4.53
Four-year colleges and universities	1,438	4.86
Two-year colleges and universities	1,438	1.78
Postsecondary Completion (%)		
Any degree, certificate, or license	1,438	3.50
Highest degree, certificate, or license earned		
Bachelor's degree or higher	1,438	3.71
Associate's degree	1,438	1.61
Certificate or license	1,438	1.53

Source: MPR analysis file "UBMSImpact6.sas."

Note: Standard errors account for project clustering and were estimated using Taylor series linearization methods.

Table D.14. Sample Sizes and Standard Errors for Reported Impact Estimates: Table C.2

	Participated in Regular Upward		Did Not Participate in Regular	
	Sample Size	Standard Error	Sample Size	Standard Error
Postsecondary Enrollment (%)				
Any postsecondary institution	552	3.46	886	1.92
Highest level of schooling attended				
Four-year college or university	552	5.18	886	2.68
Two-year college	552	4.89	886	2.07
Vocational institution	552	0.00	886	0.92
College Selectivity (%)				
Most selective four-year college or university				
More selective	552	5.18	886	3.69
Less selective	552	5.87	886	4.19
Postsecondary Credits Earned (mean)				
Two- and four-year colleges and universities	552	6.67	886	4.89
Four-year colleges and universities	552	7.39	886	5.36
Two-year colleges and universities	552	3.36	886	1.90
Postsecondary Completion (%)				
Any degree, certificate, or license	552	5.98	886	3.82
Highest degree, certificate, or license earned				
Bachelor's degree or higher	552	4.97	886	4.11
Associate's degree	552	2.77	886	1.83
Certificate or license	552	2.02	886	1.90

Source: MPR analysis file "UBMSImpact6.sas."

Note: Standard errors account for project clustering and were estimated using Taylor series linearization methods.

Table D.15. Sample Sizes and Standard Errors for Reported Impact Estimates: Table C.3

	Male		Female	
	Sample Size	Standard Error	Sample Size	Standard Error
Postsecondary Enrollment (%)				
Any postsecondary institution	480	3.07	957	2.12
Highest level of schooling attended				
Four-year college or university	480	4.09	957	3.04
Two-year college	480	3.90	957	2.09
Vocational institution	480	0.81	957	0.87
College Selectivity (%)				
Most selective four-year college or university				
More selective	480	5.11	957	3.63
Less selective	480	5.59	957	4.47
Postsecondary Credits Earned (mean)				
Two- and four-year colleges and universities	480	6.71	957	5.12
Four-year colleges and universities	480	7.10	957	5.21
Two-year colleges and universities	480	2.95	957	2.26
Postsecondary Completion (%)				
Any degree, certificate, or license	480	4.92	957	4.04
Highest degree, certificate, or license earned				
Bachelor's degree or higher	480	4.57	957	4.03
Associate's degree	480	2.87	957	1.75
Certificate or license	480	2.21	957	1.98

Source: MPR analysis file "UBMSImpact6.sas."

Note: Standard errors account for project clustering and were estimated using Taylor series linearization methods.

Table D.16. Sample Sizes and Standard Errors for Reported Impact Estimates: Table C.4

	African American		White		Hispanic	
	Sample Size	Standard Error	Sample Size	Standard Error	Sample Size	Standard Error
Postsecondary Enrollment (%)						
Any postsecondary institution	523	2.64	468	3.43	261	4.57
Highest level of schooling attended						
Four-year college or university	523	3.79	468	4.96	261	5.10
Two-year college	523	2.98	468	4.74	261	4.41
Vocational institution	523	0.96	468	1.52	261	0.00
College Selectivity (%)						
Most selective four-year college or university						
More selective	523	4.93	468	5.85	261	7.88
Less selective	523	5.90	468	6.13	261	7.49
Postsecondary Credits Earned (mean)						
Two- and four-year colleges and universities	523	7.08	468	7.12	261	10.13
Four-year colleges and universities	523	7.25	468	8.83	261	9.96
Two-year colleges and universities	523	2.10	468	4.00	261	3.97
Postsecondary Completion (%)						
Any degree, certificate, or license	523	6.33	468	4.84	261	6.69
Highest degree, certificate, or license earned						
Bachelor's degree or higher	523	6.12	468	4.94	261	6.59
Associate's degree	523	3.18	468	3.47	261	4.23
Certificate or license	523	2.70	468	0.99	261	4.00

Source: MPR analysis file "UBMSImpact6.sas."

Note: Standard errors account for project clustering and were estimated using Taylor series linearization methods.

In: Mathematics and Science Education
Editor: Chad P. Allerton
ISBN: 978-1-60692-313-9

Chapter 2

HIGHLIGHTS FROM PISA 2006: PERFORMANCE OF U.S. 15-YEAR-OLD STUDENTS IN SCIENCE AND MATHEMATICS LITERACY IN AN INTERNATIONAL CONTEXT*

Stéphane Baldi Ying Jin, Melanie Skemer, Patricia J. Green, Deborah Herget and Holly Xie

ABSTRACT

The Program for International Student Assessment (PISA) is a system of international assessments that measures 15-year-olds' performance in reading literacy, mathematics literacy, and science literacy every three years. PISA, first implemented in 2000, is sponsored by the Organization for Economic Cooperation and Development (OECD), an intergovernmental organization of 30 member countries. In 2006, 57 jurisdictions participated in PISA, including 30 OECD jurisdictions and 27 non-OECD jurisdictions.

Each PISA data collection effort assesses one of the three subject areas in depth. In this third cycle, PISA 2006, science literacy was the subject area assessed in depth. The PISA assessment measures student performance on a combined science literacy scale and on three science literacy subscales: *identifying scientific issues, explaining phenomena scientifically,* and *using scientific evidence*. Combined science literacy scores are reported on a scale from 0 to 1,000 with a mean set at 500 and a standard deviation of 100.

This report focuses on the performance of U.S. students in the major subject area of science literacy as assessed in PISA 2006.[1] Achievement in the minor subject area of mathematics literacy in 2006 is also presented.[2]

---

* Excerpted from Baldi, S., Jin, Y., Skemer, M., Green, P.J., and Herget, D. (2007). Highlights From PISA 2006: Performance of U.S. 15-Year-Old Students in Science and Mathematics Literacy in an International Context (NCES 2008–016). National Center for Education Statistics, Institute of Education Sciences, U.S. Department of Education. Washington, DC.

1 A total of 166 schools and 5,611 students participated in the assessment. The overall weighted school response rate was 69 percent before the use of replacement schools. The fi nal weighted student response rate was 91 percent.

Differences in achievement by selected student characteristics are covered in the final section.

Key findings from the report include:

- Fifteen-year-old students in the United States had an average score of 489 on the combined science literacy scale, lower than the OECD average score of 500. U.S. students scored lower on science literacy than their peers in 16 of the other 29 OECD jurisdictions and six of the 27 non-OECD jurisdictions. Twenty-two jurisdictions (five OECD jurisdictions and 17 non-OECD jurisdictions) reported lower scores compared to the United States in science literacy.
- When comparing the performance of the highest achieving students—those at the 90th percentile—there was no measurable difference between the average score of U.S. students (628) compared to the OECD average (622) on the combined science literacy scale. Twelve jurisdictions (nine OECD jurisdictions and three non-OECD jurisdictions) had students at the 90th percentile with higher scores than the United States on the combined science literacy scale.
- U.S. students also had lower scores than the OECD average score for two of the three content area subscales (*explaining phenomena scientifically* (486 versus 500) and *using scientific evidence* (489 versus 499). There was no measurable difference in the performance of U.S. students compared with the OECD average on the *identifying scientific issues* subscale (492 versus 499).
- Along with scale scores, PISA 2006 uses six proficiency levels to describe student performance in science literacy, with level 6 being the highest level of proficiency. The United States had greater percentages of students below level 1 (eight percent) and at level 1 (17 percent) than the OECD average percentages on the combined science literacy scale (five percent below level 1 and 14 percent at level 1).
- In 2006, the average U.S. score in mathematics literacy was 474, lower than the OECD average score of 498. Thirty-one jurisdictions (23 OECD jurisdictions and 8 non-OECD jurisdictions) scored higher, on average, than the United States in mathematics literacy in 2006. In contrast, 20 jurisdictions (four OECD jurisdictions and 16 non-OECD jurisdictions) scored lower than the United States in mathematics literacy in 2006.
- When comparing the performance of the highest achieving students—those at the 90th percentile—U.S. students scored lower (593) than the OECD average (615) on the mathematics literacy scale. Twenty-nine jurisdictions (23 OECD jurisdictions and six non-OECD jurisdictions) had students at the 90th percentile with higher scores than the United States on the mathematics literacy scale.
- There was no measurable difference on the combined science literacy scale between 15-year-old male (489) and female (489) students in the United States. In contrast, the OECD average was higher for males (501) than females (499) on the combined science literacy scale.
- On the combined science literacy scale, black (non-Hispanic) students (409) and Hispanic students (439) scored lower, on average, than white (non-Hispanic) students (523), Asian (non- Hispanic) students (499), and students of more than one race (non-Hispanic) (501). Hispanic students, in turn, scored higher than

[2] PISA 2006 reading literacy results are not reported for the United States because of an error in printing the test booklets. In several areas of the reading literacy assessment, students were incorrectly instructed to refer to the passage on the "opposite page" when, in fact, the necessary passage appeared on the previous page. Because of the small number of items used in assessing reading literacy, it was not possible to recalibrate the score to exclude the affected items. Furthermore, as a result of the printing error, the mean performance in mathematics and science may be misestimated by approximately 1 score point. The impact is below one standard error. For details see appendix B.

black (non- Hispanic) students, while white (non-Hispanic) students scored higher than Asian (non-Hispanic) students.

ACKNOWLEDGMENTS

This report reflects the contributions of many individuals. The authors wish to thank all those who assisted with PISA 2006, from the design stage through the creation of this report. At NCES, the project was reviewed by Eugene Owen and Marilyn Seastrom. Sampling and data collection were conducted by RTI International. The members of the PISA 2006 Expert Panel (noted in appendix D) lent their time and expertise toward reviewing the project. All data tables, figures, and text presented in the report were reviewed by Anindita Sen, Martin Hahn, Gillian Hampden-Thompson, Lydia Malley, Steve Hocker, Aparna Sundaram, and Siri Warkentien at the Education Statistics Services Institute (ESSI). We also thank our colleagues at the American Institutes for Research (AIR) who assisted with data analyses and the preparation of the report. Finally, the authors wish to thank the many principals, teachers, and students who gave generously of their time to participate in PISA 2006.

INTRODUCTION

PISA in Brief

The Program for International Student Assessment (PISA) is a system of international assessments that measures 15-year-olds' performance in reading literacy, mathematics literacy, and science literacy every 3 years. PISA was first implemented in 2000 (figure 1).

PISA is sponsored by the Organization for Economic Cooperation and Development (OECD), an intergovernmental organization of 30 member countries. In 2006, fifty-seven jurisdictions participated in PISA, including 30 OECD countries referred to throughout as jurisdictions and 27 non- OECD jurisdictions (figure 2 and table 1).

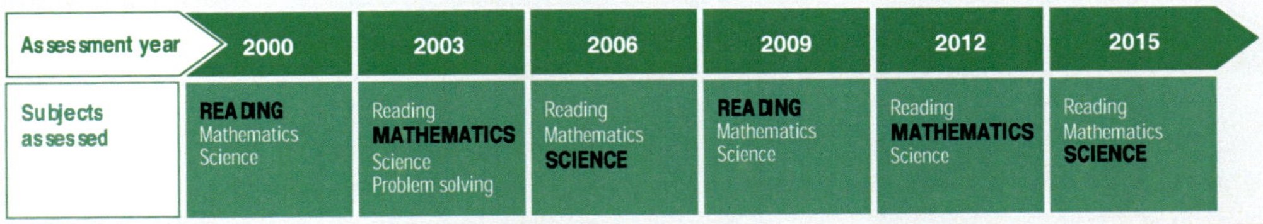

NOTE: Each subject area is tested in all assessment cycles of the Program for International Student Assessment (PISA). The subject in all capital letters is the major subject area for that cycle.
SOURCE: Organization for Economic Cooperation and Development (OECD), Program for International Student Assessment (PISA), 2006.

Figure 1. PISA administration cycle

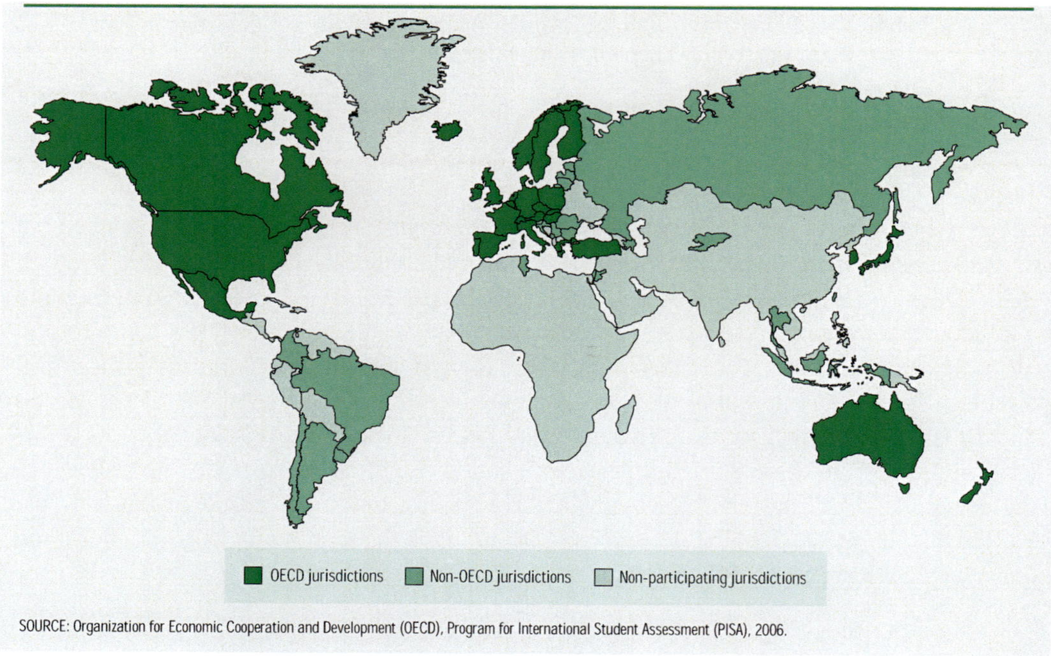

Figure 2. Jurisdictions that participated in PISA 2006

Each PISA data collection effort assesses one of the three subject areas in depth (considered the major subject area), even as all three are assessed in each cycle (the other two subjects are considered minor subject areas for that assessment year). This allows participating jurisdictions to have an ongoing source of achievement data in every subject area. In this third cycle, PISA 2006, science literacy was the subject area assessed in depth. In 2009, PISA will focus on reading literacy, which was also assessed as the major subject area in 2000.

This report focuses on the performance of U.S. students in the major subject area of science literacy as assessed in PISA 2006. Achievement in the minor subject area of mathematics literacy in 2006 is also presented,[3] as are differences in achievement by selected student characteristics.

The Unique Contribution of PISA

The United States has conducted surveys of student achievement at a variety of grade levels and in a variety of subject areas through the National Assessment of Educational Progress (NAEP) for many years. NAEP provides a regular benchmark for states and the nation and a means to monitor progress in achievement over time.

[3] PISA 2006 reading literacy results are not reported for the United States because of an error in printing the test booklets. In several areas of the reading literacy assessment, students were incorrectly instructed to refer to the passage on the "opposite page" when, in fact, the necessary passage appeared on the previous page. Because of the small number of items used in assessing reading literacy, it was not possible to recalibrate the score to exclude the affected items. Furthermore, as a result of the printing error, the mean performance in mathematics and science may be misestimated by approximately 1 score point. The impact is below one standard error. For details see appendix B.

Table 1. Participation in PISA, by jurisdiction: 2000, 2003, and 2006

Jurisdiction	2000	2003	2006
OECD jurisdictions			
Australia			
Austria			
Belgium			
Canada			
Czech Republic			
Denmark			
Finland			
France			
Germany			
Greece			
Hungary			
Iceland			
Ireland			
Italy			
Japan			
Korea, Republic of			
Luxembourg			
Mexico			
Netherlands			
New Zealand			
Norway			
Poland			
Portugal			
Slovak Republic			
Spain			
Sweden			
Switzerland			
Turkey			
United Kingdom			
United States			

Jurisdiction	2000	2003	2006
Non-OECD jurisdictions			
Argentina			
Azerbaijan			
Brazil			
Bulgaria			
Chile			
Chinese Taipei			
Colombia			
Croatia			
Estonia			
Hong Kong-China			
Indonesia			
Israel			
Jordan			
Kyrgyz Republic			
Latvia			
Liechtenstein			
Lithuania			
Macao-China			
Qatar			
Republic of Montenegro[1]			
Republic of Serbia[1]			
Romania			
Russian Federation			
Slovenia			
Thailand			
Tunisia			
Uruguay			

[1] The Republics of Montenegro and Serbia were a united jurisdiction under the PISA 2003 assessment.
NOTE: A " " indicates that the jurisdiction participated in the Program for International Student Assessment (PISA) in the speci c year. Highlighted are jurisdictions that participated in PISA in all 3 years. Because PISA is principally an Organization for Economic Cooperation and Development (OECD) study, non-OECD jurisdictions are displayed separately from the OECD jurisdictions.
SOURCE: Organization for Economic Cooperation and Development (OECD), Program for International Student Assessment (PISA), 2000, 2003, and 2006.

To provide a critical external perspective on the achievement of U.S. students through comparisons with students of other nations, the United States participates at the international level in PISA, the Progress in International Reading Literacy Study (PIRLS), and the Trends in International Mathematics and Science Study (TIMSS).[4] TIMSS and PIRLS seek to measure students' mastery of specific knowledge, skills, and concepts and are designed to reflect curriculum frameworks in the United States and other participating jurisdictions.

PISA provides a unique and complementary perspective to these studies by not focusing explicitly on curricular outcomes, but on the application of knowledge in reading, mathematics, and science to problems with a real-life context (OECD 1999). The framework for each subject area is based on concepts, processes, and situations or contexts (OECD 2006). For example, for science literacy, the concepts included are physics, chemistry, biological sciences, and earth and space sciences. The processes are centered on the ability to acquire, interpret, and act on evidence such as describing scientific phenomena and interpreting scientific evidence. The situations or contexts are those (either personal or educational) in which students might encounter scientific concepts and processes. Assessment items are then developed on the basis of these descriptions (see appendix A for examples).

[4] The United States has also participated in international comparative assessments of civics knowledge and skills (CivEd 1999) and adult literacy (International Adult Literacy Survey [IALS 1994] and Adult Literacy and Lifeskills Survey [ALL 2003]).

PISA uses the terminology of "literacy" in each subject area to denote its broad focus on the application of knowledge and skills. For example, PISA seeks to assess whether1 5-year-olds are scientifically literate, or to what extent they can apply scientific knowledge and skills to a range of different situations they may encounter in their lives. Literacy itself refers to a continuum of skills—it is not a condition that one has or does not have (i.e., literacy or illiteracy). Rather, each person's skills place that person at a particular point on the literacy continuum (OECD 2006).

The target age of 15 allows jurisdictions to compare outcomes of learning as students near the end of compulsory schooling. PISA's goal is to answer the question "what knowledge and skills do students have at age 15?" taking into account schooling and other factors that may influence their performance. In this way, PISA's achievement scores represent a "yield" of learning at age 15, rather than a direct measure of attained curriculum knowledge at a particular grade level, because 15-year-olds in the United States and elsewhere come from several grade levels (figure 3 and table C-1).

How PISA 2006 Was Conducted

PISA 2006 was sponsored by the OECD and carried out at the international level through a contract with the PISA Consortium, led by the Australian Council for Educational Research (ACER).[5] The National Center for Education Statistics (NCES) of the Institute of Education Sciences at the U.S. Department of Education was responsible for the implementation of PISA in the United States. Data collection in the United States was carried out through a contract with RTI International. An expert panel (see appendix D for a list of members) provided input on the development and dissemination of PISA in the United States.

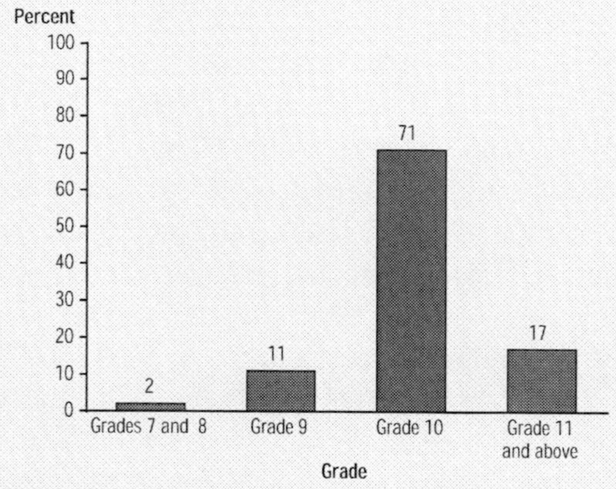

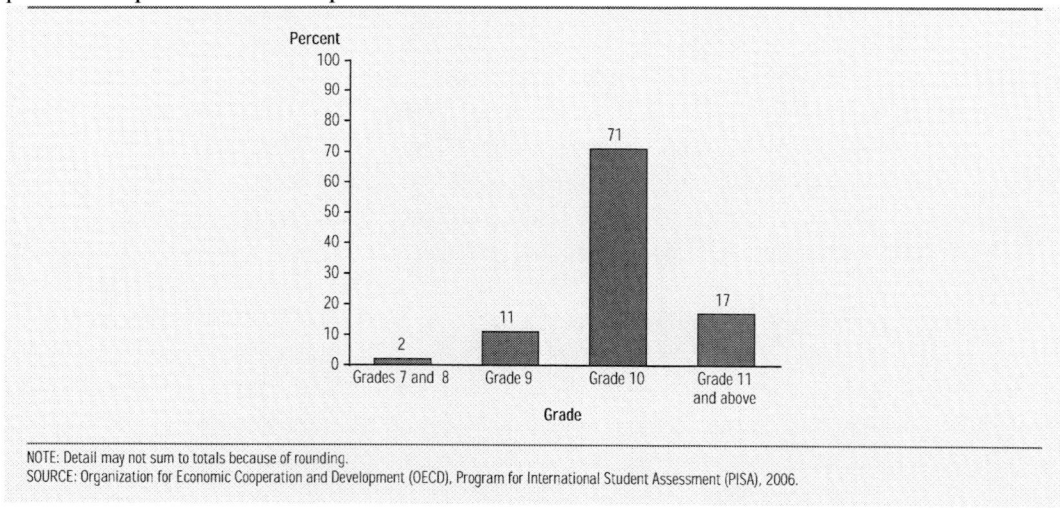

NOTE: Detail may not sum to totals because of rounding.
SOURCE: Organization for Economic Cooperation and Development (OECD), Program for International Student Assessment (PISA), 2006.

Figure 3. Percentage distribution of U.S. 15-year-old students, by grade level: 2006

[5] The PISA Consortium consists of ACER, the National Institute for Educational Policy Research (NIER, Japan), Westat (USA), the Netherlands National Institute for Educational Measurement (CITO), and the Educational Testing Service (ETS, USA).

PISA 2006 was a 2-hour paper-and-pencil assessment of 15-year-olds collected from nationally representative samples in participating jurisdictions. Like other large- scale assessments, PISA was not designed to provide individual student scores, but rather national and group estimates of performance. In PISA 2006, every student answered science items. Not every student answered both reading and mathematics items as these were distributed across different versions of the test booklets (for more information on PISA 2006's design, see the technical notes in appendix B).

PISA 2006 was administered between September and November 2006. The U.S. sample included both public and private schools, randomly selected and weighted to be representative of the nation.[6] In total, 166 schools and 5,611 students participated in PISA 2006 in the United States. The overall weighted school response rate was 69 percent before the use of replacement schools. The final weighted student response rate was 91 percent[7] (see the technical notes in appendix B for additional details on sampling, administration, response rates, and other issues).

This report provides results for the United States in relation to the other jurisdictions participating in PISA 2006, distinguishing OECD jurisdictions and non- OECD jurisdictions. All differences described in this report have been tested for statistical significance at the .05 level. Additional information on the statistical procedures used in this report is provided in the technical notes in appendix B. For further results from PISA 2006, see the OECD publication *PISA 2006: Science Competencies for Tomorrow's World* (Vols. 1 and 2) available at http://www.pisa.oecd.org (OECD, 2007a, 2007b).

PISA's major focus in 2006 was science literacy. Science literacy is defined as

> an individual's scientific knowledge and use of that knowledge to identify questions, to acquire new knowledge, to explain scientific phenomena, and to draw evidence-based conclusions about science- related issues, understanding of the characteristic features of science as a form of human knowledge and enquiry, awareness of how science and technology shape our material, intellectual, and cultural environments, and willingness to engage in science-related issues, and with the ideas of science, as a reflective citizen (OECD 2006, p.1 2).

In the PISA 2006 science literacy assessment, students completed exercises designed to assess their performance in using a range of scientific competencies, grouped and described as "competency clusters." These clusters—identifying scientific issues, explaining phenomena scientifically, using scientific evidence—describe sets of skills students may use for scientific investigation. PISA 2006 provides scores on three subscales based on these competency clusters in addition to providing a combined science literacy score.

---

[6] The sample frame data for the United States for public schools were from the 2003–04 Common Core of Data (CCD), and the data for private schools were from the 2003–04 Private School Universe Survey (PSS). Any school containing at least one 7th- through 12th-grade class as of school year 2003–04 was included in the school sampling frame.

[7] Response rates reported here are based on the formula used in the international report and are not consistent with NCES standards. A more conservative way to calculate the response rate would be to include replacement schools that participated in the denominator as well as the numerator, and to add replacement schools that were hard refusals to the denominator. This results in a response rate of 67.5 percent.

- **Identifying scientific issues** includes recognizing issues that are possible to investigate scientifically; identifying keywords to search for scientific information; and recognizing the key features of a scientific investigation.
- **Explaining phenomena scientifically** covers applying knowledge of science in a given situation; describing or interpreting phenomena scientifically and predicting changes; and identifying appropriate descriptions, explanations, and predictions.
- **Using scientific evidence** includes interpreting scientific evidence and making and communicating conclusions; identifying the assumptions, evidence, and reasoning behind conclusions; and reflecting on the societal implications of science and technological developments.

Sample science literacy items (and examples of student responses for each item) for each competency cluster are shown in appendix A.

Combined science literacy scores are reported on a scale from 0 to 1,000 with a mean set at 500 and a standard deviation of 100.[8] Fifteen-year-old students in the United States had an average score of 489 on the combined science literacy scale, lower than the OECD average score of 500 (tables 2 and C-2). U.S. students scored lower in science literacy than their peers in 16 of the other 29 OECD jurisdictions and 6 of the 27 non- OECD jurisdictions. Twenty-two jurisdictions (5 OECD jurisdictions and 17 non-OECD jurisdictions) reported lower scores than the United States in science literacy.

When comparing the performance of the highest achieving students—those at the 90th percentile—there was no measurable difference between the average score of U.S. students (628) compared to the OECD average (622) on the combined science literacy scale (table C-3). Twelve jurisdictions (9 OECD jurisdictions and 3 non-OECD jurisdictions) had students at the 90th percentile with higher scores than the United States on the combined science literacy scale.

At the other end of the distribution, among low- achieving students at the 10th percentile, U.S. students scored lower (349) than the OECD average (375) on the combined science literacy scale. Thirty jurisdictions (21 OECD jurisdictions and 9 non-OECD jurisdictions) had students at the 10th percentile with higher scores than the United States on the combined science literacy scale.

U.S. students also had lower scores than the OECD average score for two of the three scientific literacy subscales (*explaining phenomena scientifically* (486 versus 500) and *using scientific evidence* (489 versus 499)). Twenty-five jurisdictions (19 OECD and 6 non-OECD jurisdictions) had a higher average score than the United States on the *explaining phenomena scientifically* subscale, and 20 jurisdictions (14 OECD and 6 non-OECD jurisdictions) had a higher average score than the United States on the *using scientific evidence* subscale. There was no measurable difference in the performance of U.S. students compared with the OECD average on the *identifying scientific issues* subscale (492 versus 499). However, 18 jurisdictions (13 OECD and 5 non-OECD jurisdictions) scored higher than the United States on the *identifying scientific issues* subscale.

[8] The combined science literacy scale is made up of all items in the three subscales. However, the combined science scale and the three subscales are each computed separately through Item Response Theory (IRT) models. Therefore, the combined science scale score is not the average of the three subscale scores. For details on the computation of the science literacy scale and subscales see Adams (in press).

Table 2. Average scores of 15-year-old students on combined science literacy scale and science literacy subscales, by jurisdiction: 2006

Combined science literacy scale

Jurisdiction	Score
OECD average	500
OECD jurisdictions	
Finland	563
Canada	534
Japan	531
New Zealand	530
Australia	527
Netherlands	525
Korea, Republic of	522
Germany	516
United Kingdom	515
Czech Republic	513
Switzerland	512
Austria	511
Belgium	510
Ireland	508
Hungary	504
Sweden	503
Poland	498
Denmark	496
France	495
Iceland	491
United States	**489**
Slovak Republic	488
Spain	488
Norway	487
Luxembourg	486
Italy	475
Portugal	474
Greece	473
Turkey	424
Mexico	410
Non-OECD jurisdictions	
Hong Kong-China	542
Chinese Taipei	532
Estonia	531
Liechtenstein	522
Slovenia	519
Macao-China	511
Croatia	493
Latvia	490
Lithuania	488
Russian Federation	479
Israel	454
Chile	438
Republic of Serbia	436
Bulgaria	434
Uruguay	428
Jordan	422
Thailand	421
Romania	418
Republic of Montenegro	412
Indonesia	393
Argentina	391
Brazil	390
Colombia	388
Tunisia	386
Azerbaijan	382
Qatar	349
Kyrgyz Republic	322

Science literacy subscales

Identifying scientiÞc issues

Jurisdiction	Score
OECD average	499
OECD jurisdictions	
Finland	555
New Zealand	536
Australia	535
Netherlands	533
Canada	532
Japan	522
Korea, Republic of	519
Ireland	516
Belgium	515
Switzerland	515
United Kingdom	514
Germany	510
Austria	505
Czech Republic	500
France	499
Sweden	499
Iceland	494
Denmark	493
United States	**492**
Norway	489
Spain	489
Portugal	486
Poland	483
Luxembourg	483
Hungary	483
Slovak Republic	475
Italy	474
Greece	469
Turkey	427
Mexico	421
Non-OECD jurisdictions	
Hong Kong-China	528
Liechtenstein	522
Slovenia	517
Estonia	516
Chinese Taipei	509
Croatia	494
Macao-China	490
Latvia	489
Lithuania	476
Russian Federation	463
Israel	457
Chile	444
Republic of Serbia	431
Uruguay	429
Bulgaria	427
Thailand	413
Romania	409
Jordan	409
Colombia	402
Republic of Montenegro	401
Brazil	398
Argentina	395
Indonesia	393
Tunisia	384
Azerbaijan	353
Qatar	352
Kyrgyz Republic	321

Explaining phenomena scientiÞcally

Jurisdiction	Score
OECD average	500
OECD jurisdictions	
Finland	566
Canada	531
Czech Republic	527
Japan	527
New Zealand	522
Netherlands	522
Australia	520
Germany	519
Hungary	518
United Kingdom	517
Austria	516
Korea, Republic of	512
Sweden	510
Switzerland	508
Poland	506
Ireland	505
Belgium	503
Denmark	501
Slovak Republic	501
Norway	495
Spain	490
Iceland	488
United States	**486**
Luxembourg	483
France	481
Italy	480
Greece	476
Portugal	469
Turkey	423
Mexico	406
Non-OECD jurisdictions	
Hong Kong-China	549
Chinese Taipei	545
Estonia	541
Slovenia	523
Macao-China	520
Liechtenstein	516
Lithuania	494
Croatia	492
Latvia	486
Russian Federation	483
Bulgaria	444
Israel	443
Republic of Serbia	441
Jordan	438
Chile	432
Romania	426
Uruguay	423
Thailand	420
Republic of Montenegro	417
Azerbaijan	412
Indonesia	395
Brazil	390
Argentina	386
Tunisia	383
Colombia	379
Qatar	356
Kyrgyz Republic	334

Using scientiÞc evidence

Jurisdiction	Score
OECD average	499
OECD jurisdictions	
Finland	567
Japan	544
Canada	542
Korea, Republic of	538
New Zealand	537
Australia	531
Netherlands	526
Switzerland	519
Belgium	516
Germany	515
United Kingdom	514
France	511
Ireland	506
Austria	505
Czech Republic	501
Hungary	497
Sweden	496
Poland	494
Luxembourg	492
Iceland	491
Denmark	489
United States	**489**
Spain	485
Slovak Republic	478
Norway	473
Portugal	472
Italy	467
Greece	465
Turkey	417
Mexico	402
Non-OECD jurisdictions	
Hong Kong-China	542
Liechtenstein	535
Chinese Taipei	532
Estonia	531
Slovenia	516
Macao-China	512
Latvia	491
Croatia	490
Lithuania	487
Russian Federation	481
Israel	460
Chile	440
Uruguay	429
Republic of Serbia	425
Thailand	423
Bulgaria	417
Romania	407
Republic of Montenegro	407
Jordan	405
Indonesia	386
Argentina	385
Colombia	383
Tunisia	382
Brazil	378
Azerbaijan	344
Qatar	324
Kyrgyz Republic	288

■ Average is higher than the U.S. average   □ Average is not measurably different from the U.S. average   □ Average is lower than the U.S. average

NOTE: The Organization for Economic Cooperation and Development (OECD) average is the average of the national averages of the OECD member jurisdictions. Because the Program for International Student Assessment (PISA) is principally an OECD study, the results for non-OECD jurisdictions are displayed separately from those of the OECD jurisdictions and are not included in the OECD average. Jurisdictions are ordered on the basis of average scores, from highest to lowest within the OECD jurisdictions and non-OECD jurisdictions. Combined science literacy scores are reported on a scale from 0 to 1,000. Because of an error in printing the test booklets, the United States mean performance may be misestimated by approximately 1 score point. The impact is below one standard error. For details see appendix B. Score differences as noted between the United States and other jurisdictions (as well as between the United States and the OECD average) are signiÞcantly different at the .05 level of statistical signiÞcance.
SOURCE: Organization for Economic Cooperation and Development (OECD), Program for International Student Assessment (PISA), 2006.

Exhibit 1. Description of general competencies and tasks students should be able to do, by proficiency level for the combined science literacy scale: 2006

Proficiency level	Task descriptions
Level 1	At level 1, students have such a limited scientific knowledge that it can only be applied to a few familiar situations. They should be able to present scientific explanations that are obvious and follow concretely from given evidence.
Level 2	At level 2, students have adequate scientific knowledge to provide possible explanations in familiar contexts or draw conclusions based on simple investigations. They should be capable of direct reasoning and making literal interpretations of the results of scientific inquiry or technological problem solving.
Level 3	At level 3, students should be able to identify clearly described scientific issues in a range of contexts. They should be able to select facts and knowledge to explain phenomena and apply simple models or inquiry strategies. Students at this level should be able to interpret and use scientific concepts from different disciplines and apply them directly. They should be able to develop short communications using facts and make decisions based on scientific knowledge.
Level 4	At level 4, students should be able to work effectively with situations and issues that may involve explicit phenomena requiring them to make inferences about the role of science or technology. They should be able to select and integrate explanations from different disciplines of science or technology and link those explanations directly to aspects of life situations. Students at this level should be able to reflect on their actions and communicate decisions using scientific knowledge and evidence.
Level 5	At level 5, students should be able to identify the scientific components of many complex life situations; apply both scientific concepts and knowledge about science to these situations; and should be able to compare, select, and evaluate appropriate scientific evidence for responding to life situations. Students at this level should be able to use well-developed inquiry abilities, link knowledge appropriately, and bring critical insights to these situations. They should be able to construct evidence-based explanations and arguments based on their critical analysis.
Level 6	At level 6, students should be able to consistently identify, explain, and apply scientific knowledge and knowledge about science in a variety of complex life situations. They should be able to link different information sources and explanations and use evidence from those sources to justify decisions. They should be able to clearly and consistently demonstrate advanced scientific thinking and reasoning, and they are willing to use their scientific understanding in support of solutions to unfamiliar scientific and technological situations. Students at this level should be able to use scientific knowledge and develop arguments in support of recommendations and decisions that center on personal, social, or global situations.

Note: To reach a particular proficiency level, a student must correctly answer a majority of items at that level. Students were classified into science literacy levels according to their scores. Exact cut point scores are as follows: below level 1 (a score less than or equal to 334.94); level 1 (a score greater than 334.94 and less than or equal to 409.54); level 2 (a score greater than 409.54 and less than or equal to 484.14); level 3 (a score greater than 484.14 and less than or equal to 558.73); level 4 (a score greater than 558.73 and less than or equal to 633.33); level 5 (a score greater than 633.33 and less than or equal to 707.93); and level 6 (a score greater than 707.93).

Source: Organization for Economic Cooperation and Development (OECD). (2006). *Assessing Scientific, Reading and Mathematical Literacy: A Framework for PISA 2006*. Paris: Author; Organization for Economic Cooperation and Development (OECD), Program for International Student Assessment (PISA), 2006.

Along with scale scores, PISA 2006 also uses six proficiency levels (levels 1 through 6, with level 6 being the highest level of proficiency) to describe student performance in science literacy (see exhibit 1 for descriptions of the proficiency levels). An additional level (below level 1) encompasses students whose skills cannot be described using these proficiency levels. The proficiency levels describe what students at each level should be able to do and allow comparisons of the percentages of students in each jurisdiction who perform at different levels of science literacy (see the technical notes in appendix B for more information about how levels were set).

The United States had greater percentages of students at or below level 1 than the OECD average percentages (figure 4, table C-5) on the combined science literacy scale. The United States also had lower percentages of students at levels 3 and 4 than the OECD average percentages. The percentages of U.S. students performing at levels 2, 5, and 6 were not measurably different from the OECD averages.

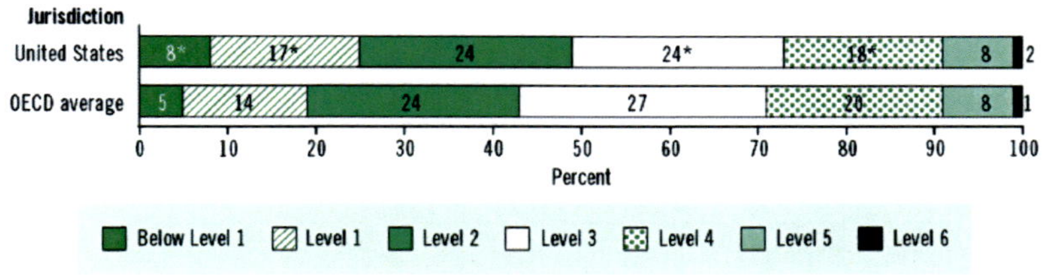

* $p < .05$. Significantly different from the corresponding OECD average percentage at the .05 level of statistical significance.

NOTE: To reach a particular proficiency level, a student must correctly answer a majority of items at that level. Students were classified into science literacy levels according to their scores. Exact cut point scores are as follows: below level 1 (a score less than or equal to 334.94); level 1 (a score greater than 334.94 and less than or equal to 409.54); level 2 (a score greater than 409.54 and less than or equal to 484.14); level 3 (a score greater than 484.14 and less than or equal to 558.73); level 4 (a score greater than 558.73 and less than or equal to 633.33); level 5 (a score greater than 633.33 and less than or equal to 707.93); and level 6 (a score greater than 707.93). The Organization for Economic Cooperation and Development (OECD) average is the average of the national averages of the OECD member jurisdictions. Because of an error in printing the test booklets, the United States mean performance may be misestimated by approximately 1 score point. The impact is below one standard error. For details see appendix B. Detail may not sum to totals because of rounding.

SOURCE: Organization for Economic Cooperation and Development (OECD), Program for International Student Assessment (PISA), 2006.

Figure 4. Percentage distribution of 15-year-old students in the United States and OECD jurisdictions on combined science literacy scale, by proficiency level: 2006

In combined science literacy in 2006, six of the other 56 jurisdictions (Australia, Canada, Finland, Japan, New Zealand, and the United Kingdom—all OECD jurisdictions) had a higher percentage of students at level 6 than the United States (figure 5, table C-5). In contrast, 19 jurisdictions had a higher percentage of students below level 1 than the United

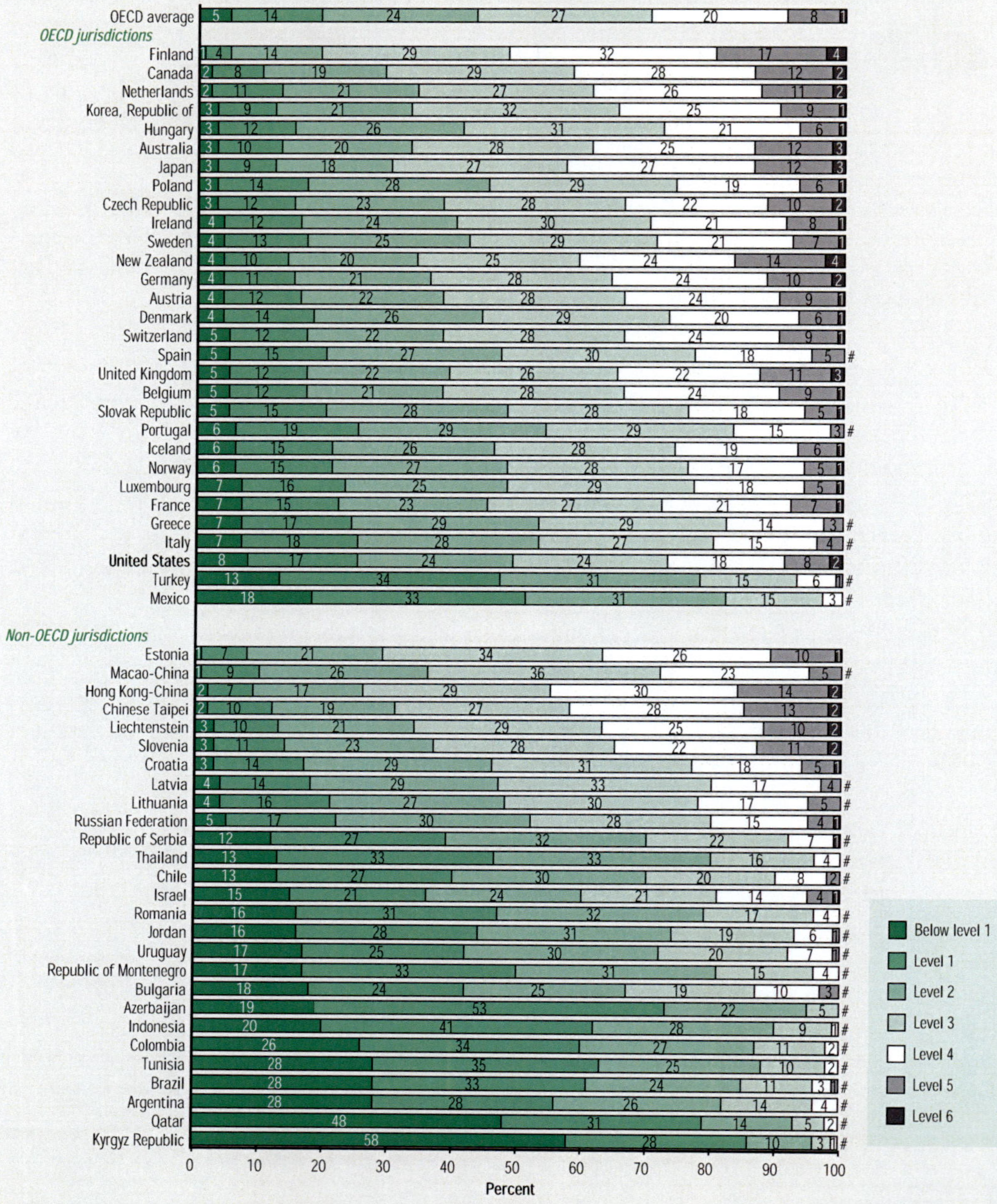

Rounds to zero.

NOTE: To reach a particular proficiency level, a student must correctly answer a majority of items at that level. Students were classified into science literacy levels according to their scores. Exact cut point scores are as follows: below level 1 (a score less than or equal to 334.94); level 1 (a score greater than 334.94 and less than or equal to 409.54); level 2 (a score greater than 409.54 and less than or equal to 484.14); level 3 (a score greater than 484.14 and less than or equal to 558.73); level 4 (a score greater than 558.73 and less than or equal to 633.33); level 5 (a score greater than 633.33 and less than or equal to 707.93); and level 6 (a score greater than 707.93). The Organization for Economic Cooperation and Development (OECD) average is the average of the national averages of the OECD member jurisdictions. Because the Program for International Student Assessment (PISA) is principally an OECD study, the results for non-OECD jurisdictions are displayed separately from those of the OECD jurisdictions and are not included in the OECD average. Jurisdictions are ordered on the basis of percentages below level 1, from lowest to highest within the OECD jurisdictions and non-OECD jurisdictions. Because of an error in printing the test booklets, the United States mean performance may be misestimated by approximately 1 score point. The impact is below one standard error. For details see appendix B. Detail may not sum to totals because of rounding.

SOURCE: Organization for Economic Cooperation and Development (OECD), Program for International Student Assessment (PISA), 2006.

Figure 5. Percentage distribution of 15-year-old students on combined science literacy scale, by proficiency level and jurisdiction: 2006

States (2 of these—Mexico and Turkey—were OECD jurisdictions). Nineteen jurisdictions (the same 2 OECD jurisdictions and 17 non-OECD jurisdictions) also had a higher percentage of students at level 1 than the United States.

U.S. PERFORMANCE IN MATHEMATICS LITERACY

In PISA 2006, mathematics literacy is defined as

> an individual's capacity to identify and understand the role that mathematics plays in the world, to make well-founded judgments and to use and engage with mathematics in ways that meet the needs of that individual's life as a constructive, concerned and reflective citizen (OECD, 2006, p.1 2).

In 2006, the average U.S. score in mathematics literacy was 474 on a scale from 0 to 1,000, lower than the OECD average score of 498 (tables 3 and C-7). Thirty-one jurisdictions (23 OECD jurisdictions and eight non-OECD jurisdictions) had a higher average score than the United States in mathematics literacy in 2006. In contrast, 20 jurisdictions (four OECD jurisdictions and 16 non-OECD jurisdictions) scored lower than the United States in mathematics literacy in 2006.

When comparing the performance of the highest achieving students—those at the 90th percentile—U.S. students scored lower (593) than the OECD average (615) on the mathematics literacy scale (table C-8). Twenty-nine jurisdictions (23 OECD jurisdictions and six non-OECD jurisdictions) had students at the 90th percentile with higher scores than the United States on the mathematics literacy scale.At the other end of the distribution, among low- achieving students at the 10th percentile, U.S. students scored lower (358) than the OECD average (379) on the mathematics literacy scale. Twenty-six jurisdictions (18 OECD jurisdictions and eight non-OECD jurisdictions) had students at the 10th percentile with higher scores than the United States on the mathematics literacy scale.

There was no measurable change in either the U.S. mathematics literacy score from 2003 to 2006 (483 versus 474) or the U.S. position compared to the OECD average, although scores in 11 other jurisdictions did change (table C-7). Four jurisdictions saw their average mathematics literacy scores increase (two non-OECD jurisdictions, Brazil and Indonesia, and two OECD jurisdictions, Greece and Mexico). The United States scored higher than all four of these jurisdictions in both 2003 and 2006. Seven jurisdictions' scores (including six OECD jurisdictions) were lower in 2006 than 2003 in mathematics literacy, although the U.S. position compared to these seven jurisdictions did not change between 2003 and 2006.

Table 3. Average scores of 15-year-old students on mathematics literacy scale, by jurisdiction: 2006

Mathematics literacy scale	
Jurisdiction	Score
OECD average	498
OECD jurisdictions	
Finland	548
Korea, Republic of	547
Netherlands	531
Switzerland	530
Canada	527
Japan	523
New Zealand	522
Belgium	520
Australia	520
Denmark	513
Czech Republic	510
Iceland	506
Austria	505
Germany	504
Sweden	502
Ireland	501
France	496
United Kingdom	495
Poland	495
Slovak Republic	492
Hungary	491
Luxembourg	490
Norway	490
Spain	480
United States	**474**
Portugal	466
Italy	462
Greece	459
Turkey	424
Mexico	406
Non-OECD jurisdictions	
Chinese Taipei	549
Hong Kong-China	547
Macao-China	525
Liechtenstein	525
Estonia	515
Slovenia	504
Lithuania	486
Latvia	486
Azerbaijan	476
Russian Federation	476
Croatia	467
Israel	442
Republic of Serbia	435
Uruguay	427
Thailand	417
Romania	415
Bulgaria	413
Chile	411
Republic of Montenegro	399
Indonesia	391
Jordan	384
Argentina	381
Colombia	370
Brazil	370
Tunisia	365
Qatar	318
Kyrgyz Republic	311

■ Average is higher than the U.S. average

□ Average is not measurably different from the U.S. average

□ Average is lower than the U.S. average

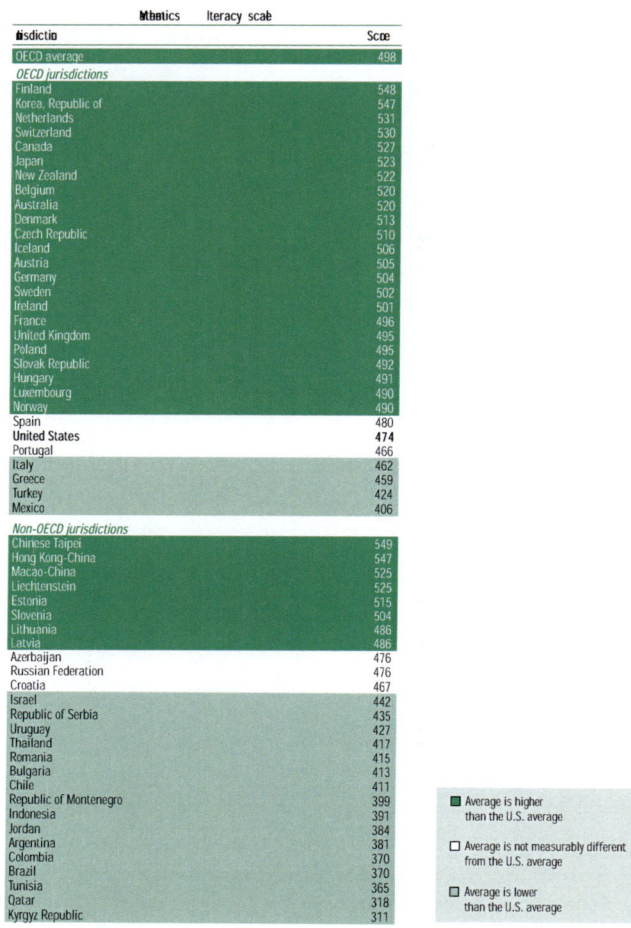

NOTE: The Organization for Economic Cooperation and Development (OECD) average is the average of the national averages of the OECD member jurisdictions. Because the Program for International Student Assessment (PISA) is principally an OECD study, the results for non-OECD jurisdictions are displayed separately from those of the OECD jurisdictions and are not included in the OECD average. Jurisdictions are ordered on the basis of average scores, from highest to lowest within the OECD jurisdictions and non-OECD jurisdictions. Mathematics literacy scores are reported on a scale from 0 to 1,000. Because of an error in printing the test booklets, the United States mean performance may be misestimated by approximately 1 score point. The impact is below one standard error. For details see appendix B. Score differences as noted between the United States and other jurisdictions (as well as between the United States and the OECD average) are signi cantly different at the .05 level of statistical signi cance.
SOURCE: Organization for Economic Cooperation and Development (OECD), Program for International Student Assessment (PISA), 2006.

Differences in Performance by Selected Student Characteristics

This section provides information about student performance on PISA 2006 by various characteristics (sex and racial/ethnic background). Because PISA 2006's emphasis was on science literacy, the focus in this section is on performance in this area. The results cannot be

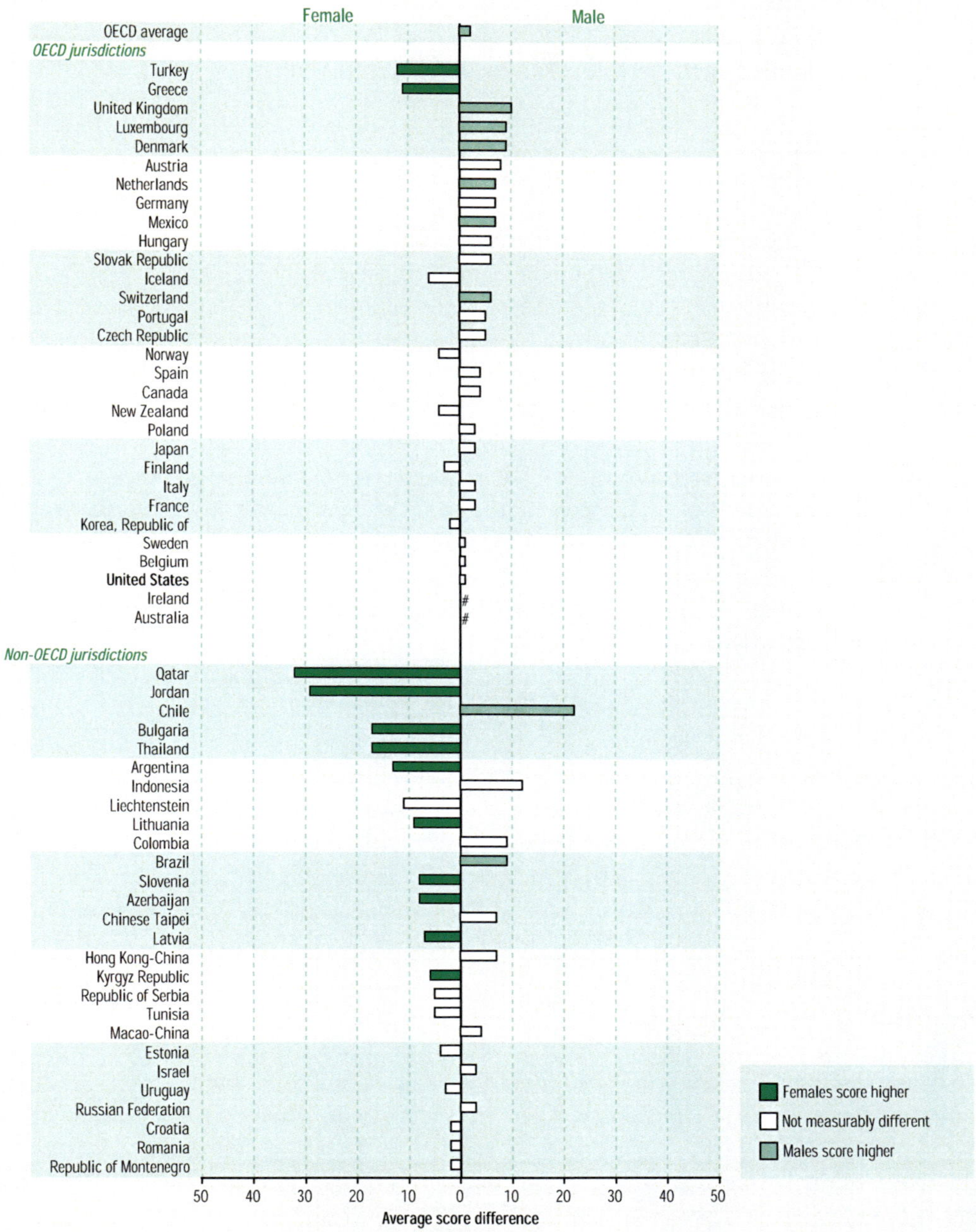

Rounds to zero.

NOTE: Each bar above represents the average score difference between males and females on the combined science literacy scale. The Organization for Economic Cooperation and Development (OECD) average is the average of the national averages of the OECD member jurisdictions. Because the Program for International Student Assessment (PISA) is principally an OECD study, the results for non-OECD jurisdictions are displayed separately from those of OECD jurisdictions and are not included in the OECD average. Jurisdictions are ordered on the basis of score differences between males and females, from largest to smallest within the OECD jurisdictions and non-OECD jurisdictions. Because of an error in printing the test booklets, the United States mean performance may be misestimated by approximately 1 score point. The impact is below one standard error. For details see appendix B. Score differences between males and females are statistically signiÞcant at the .05 level of signiÞcance.

SOURCE: Organization for Economic Cooperation and Development (OECD), Program for International Student Assessment (PISA), 2006.

Figure 6. Difference in average scores between 15-year-old male and female students on combined science literacy scale, by jurisdiction: 2006 used to establish a cause-and-effect relationship between being a member of a group and achievement in PISA 2006. Student performance can be affected by a complex mix of educational and other factors that are not examined here.

Sex

In the United States, no measurable difference was observed between the scores for 15-year-old males (489) and females (489) on the combined science literacy scale (figure 6, table C-9). Males had a higher average score than females in eight jurisdictions (six OECD jurisdictions and two non-OECD jurisdictions), while females had a higher average score than males in 12 jurisdictions (2 OECD jurisdictions and 10 non- OECD jurisdictions). The OECD average was higher for males (501) than females (499) on the combined science literacy scale.

In the United States, no measurable difference was found in the percentage of U.S. females (1.5 percent) and males (1.6 percent) scoring at level 6 (the highest level) on the combined science literacy scale (table C-1 0). Again, the percentages of U.S. females scoring at (16.2 percent) or below (6.8 percent) level 1 (the lowest levels) did not measurably differ from those for their male peers (8.3 percent below level 1 and 17.4 percent at level 1) on the combined science literacy scale.

On average across the OECD jurisdictions, females scored higher than males on the *identifying scientific issues* subscale (508 versus 490) and the *using scientific evidence* subscale (501 versus 498), while males scored higher than females on the *explaining phenomena scientifically* subscale (508 versus 493) (table C-11). In the United States, females had a higher average score than males on the *identifying scientific issues* subscale (500 versus 484), while males had a higher average score than females on the *explaining phenomena scientifically* subscale (492 versus 480).[9] There was no measurable difference between U.S. 15-year-old males and females on the *using scientific evidence* subscale (486 versus 491).

Race/Ethnicity

Racial and ethnic groups vary by country, so it is not possible to compare their performance internationally. Thus, this section refers only to the 2006 findings for the United States.

On the combined science literacy scale, black (non-Hispanic) students and Hispanic students scored lower, on average, than white (non-Hispanic) students, Asian (non-Hispanic) students, and students of more than one race (non-Hispanic) (figure 7, table C-12).[10] On average, Hispanic students scored higher than black (non-Hispanic) students, while white

[9] The effect size of the difference between two means can be calculated by dividing the raw difference in means by the pooled standard deviation of the comparison groups (see appendix B for an explanation). The effect size of the difference in achievement on the *identifying scientific issues* subscale between U.S. 15-year-old male and female students in 2006 was -.16. The effect size of the difference in achievement on the *explaining phenomena scientifically* subscale between U.S. 15-year-old male and female students in 2006 was .12.

[10] The effect size of the difference in achievement on the combined science literacy scale between White and Black and between White and Hispanic 15-year-old students in 2006 was 1.23 and .88, respectively.

(non-Hispanic) students scored higher than Asian (non-Hispanic) students. This pattern of performance on PISA 2006 by race/ethnicity is similar to that found in PISA 2000 and PISA 2003 (Lemke et al. 2001, 2004).

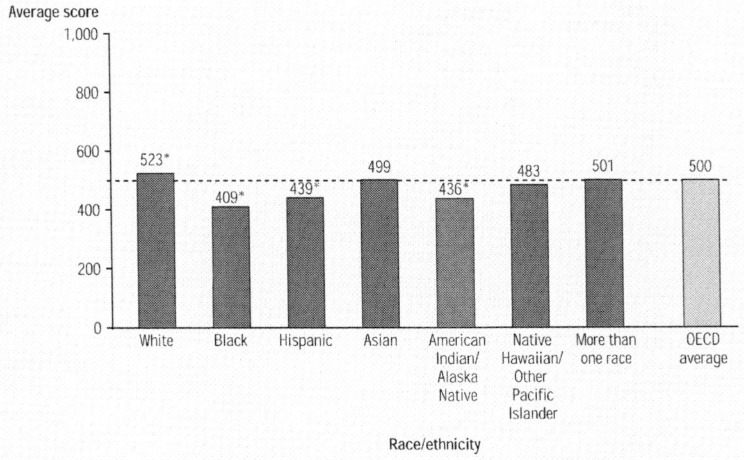

* $p < .05$. SigniÞcantly different from the OECD average at the .05 level of statistical signiÞcance.
NOTE: Black includes African American, and Hispanic includes Latino. Students who identiÞed themselves as being of Hispanic origin were classiÞed as Hispanic, regardless of their race. The Organization for Economic Cooperation and Development (OECD) average is the average of the national averages of the OECD member jurisdictions. Because of an error in printing the test booklets, the United States mean performance may be misestimated by approximately 1 score point. The impact is below one standard error. For details see appendix B.
SOURCE: Organization for Economic Cooperation and Development (OECD), Program for International Student Assessment (PISA), 2006.

Figure 7. Average scores of U.S. 15-year-old students on combined science literacy scale, by race/ethnicity: 2006

On the combined science literacy scales, black (non-Hispanic) students, Hispanic students, and American Indian/Alaska Native (non-Hispanic) students scored below the OECD average, while scores for white (non-Hispanic) students were above the OECD average. On average, the mean scores of white (non- Hispanic), Asian (non-Hispanic), and students of more than one race (non-Hispanic) were in the PISA level 3 proficiency range for the combined science literacy scale; the mean scores of Hispanic, American Indian/Alaska Native (non-Hispanic), and Native Hawaiian/Other Pacific Islander (non-Hispanic) students were in the level 2 proficiency range; and the mean score for black (non-Hispanic) students was at the top of the level 1 proficiency range.[11]

[11] To reach a particular proficiency level, a student must correctly answer a majority of items at that level. Students were classifi ed into science literacy levels according to their scores. Exact cut point scores are as follows: below level 1 (a score less than or equal to 334.94); level 1 (a score greater than 334.94 and less than or equal to 409.54); level 2 (a score greater than 409.54 and less than or equal to 484.14); level 3 (a score greater than 484.14 and less than or equal to 558.73); level 4 (a score greater than 558.73 and less than or equal to 633.33); level 5 (a score greater than 633.33 and less than or equal to 707.93); and level 6 (a score greater than 707.93).

For Further Information

This chapter provides selected findings from PISA 2006 from a U.S. perspective. Readers may be interested in exploring other aspects of PISA's results. Additional findings are presented in the OECD report, *PISA 2006: Science Competencies for Tomorrow's World* (Vols. 1 and 2), which can be found at http://www.pisa.oecd.org (OECD, 2007a, 2007b). Data with which researchers can conduct their own analyses are also available at this site.

References

Adams, R. (Ed.). (in press). *PISA 2006 Technical Report*. Paris: Organization for Economic Cooperation and Development.

Adams, R. (Ed.). (2004). *PISA 2003 Technical Report*. Paris: Organization for Economic Cooperation and Development.

Australian Council for Educational Research (ACER). (2006). *PISA 2006 Main Study Management Manual, Version 3*. Camberwell, Australia: Author.

Cohen, J. (1988). *Statistical Power Analysis for the Behavioral Sciences* (2nd ed.). Hillsdale, NJ: Lawrence Erlbaum Associates.

Krotki, K. and Bland, C. (2008). *Program for International Student Assessment (PISA) 2006 Nonresponse Bias Analysis for the United States*. (NCE S 2008-018). National Center for Education Statistics, Institute of Education Sciences, U.S. Department of Education, Washington, DC.

Lemke, M., Calsyn, C., Lippman, L., Jocelyn, L., Kastberg, D., Liu, Y.Y., Roey, S., Williams, T., Kruger, T., and Bairu, G. (2001). *Outcomes of Learning: Results From the 2000 Program for International Student Assessment of 15-Year-Olds in Reading, Mathematics, and Science Literacy* (NCES 2002- 115). National Center for Education Statistics, U.S. Department of Education. Washington, DC.

Lemke, M., Sen, A., Pahlke, E., Partelow, L., Miller, D., Williams, T., Kastberg, D., and Jocelyn, L. (2004). *International Outcomes of Learning in Mathematics Literacy and Problem Solving: PISA 2003 Results From the U.S. Perspective* (NCES 2005-003). National Center for Education Statistics, Institute of Education Sciences, U.S. Department of Education. Washington, DC.

Mislevy, R.J. (1988). Randomization Based Inferences About Latent Variables From Complex Samples. *Psychometrika, 56*(2): 177–196.

Organization for Economic Cooperation and Development (OECD). (1999). *Measuring Student Knowledge and Skills: A New Framework for Assessment*. Paris: Author.

Organization for Economic Cooperation and Development (OECD). (2006). *Assessing Scientific, Reading and Mathematical Literacy: A Framework for PISA 2006*. Paris: Author.

Organization for Economic Cooperation and Development (OECD). (2007a). *PISA 2006: Science Competencies for Tomorrow's World: Vol.*

Organization for Economic Cooperation and Development (OECD). (2007b). *PISA 2006: Science Competencies for Tomorrow's World: Vol.*

Rosnow, R.L., and Rosenthal, R. (1996). Computing Contrasts, Effect Sizes, and Counternulls on Other People's Published Data: General Procedures for Research Consumers. *Psychological Methods, 1*: 33 1–340.

Exhibit A-1. Map of selected science items in PISA 2006

			Competency		
			Identifying scientific issues	**Explaining phenomena scientifically**	**Using scientific evidence**
Knowledge	**Knowledge of science** (scientific content)	**Physical systems**		Acid Rain Q1 (506)	Acid Rain Q2 (460)
		Living systems			
		Earth and space systems		Grand Canyon Q2 (451) Grand Canyon Q3 (411)	
		Technology systems			
	Knowledge about science (scientific process)	**Scientific inquiry**	Acid Rain Q3 (513) (partial credit) Acid Rain Q3 (717) (full credit) Sunscreens Q1 (588) Sunscreens Q2 (499) Sunscreens Q3 (574) Grand Canyon Q1 (485)		
		Scientific explanation			Sunscreens Q4 (616) (partial credit) Sunscreens Q4 (629) (full credit)

Note: Numbers in parentheses refer to the score or proficiency level associated with the item. To reach a particular proficiency level, a student must correctly answer a majority of items at that level. Students were classified into science literacy levels according to their scores. Exact cut point scores are as follows: below level 1 (a score less than or equal to 334.94); level 1 (a score greater than 334.94 and less than or equal to 409.54); level 2 (a score greater than 409.54 and less than or equal to 484.14); level 3 (a score greater than 484.14 and less than or equal to 558.73); level 4 (a score greater than 558.73 and less than or equal to 633.33); level 5 (a score greater than 633.33 and less than or equal to 707.93); and level 6 (a score greater than 707.93).

Source: Organization for Economic Cooperation and Development (OECD). (2006). *Assessing Scientific, Reading and Mathematical Literacy: A Framework for PISA 2006*. Paris: Author.

Appendix A: Sample Science Items from PISA 2006

This section presents sample items used in the PISA 2006 science assessment. These items serve to illustrate the various competencies and types of scientific knowledge measured by PISA, as well as the different difficulty levels at which students were tested. For more information about the science literacy subject area or additional examples of science literacy items, refer to *Assessing Scientific, Reading and Mathematical Literacy: A Framework for PISA 2006* (OECD 2006).

Exhibit A-1 summarizes the distribution of the sample items across the PISA knowledge areas and competency types, along with their associated difficulty. Grand Canyon question 3, for example, tests student knowledge of science in earth and space systems under the *explaining phenomena scientifically* competency. This question has a difficulty of 41 1 (level

2) on the combined science literacy scale, requiring students to know that fossils from organisms that lived long ago may be exposed when sea levels recede.

Acid Rain

Below is a photo of statues called Caryatids that were built on the Acropolis in Athens more than 2500 years ago. The statues are made of a type of rock called marble. Marble is composed of calcium carbonate.

In 1980, the original statues were transferred inside the museum of the Acropolis and were replaced by replicas. The original statues were being eaten away by acid rain.

Question 1. Acid Rain

Normal rain is slightly acidic because it has absorbed some carbon dioxide from the air. Acid rain is more acidic than normal rain because it has absorbed gases like sulfur oxides and nitrogen oxides as well.

Where do these sulfur oxides and nitrogen oxides in the air come from?

Sulfur oxides and nitrogen oxides are put in the air from pollution and burning fossil fuels. (full credit)

Sulfur oxides and nitrogen oxides come from the pollution in the air. (partial credit)

The effect of acid rain on marble can be modeled by placing chips of marble in vinegar overnight. Vinegar and acid rain have about the same acidity level. When a marble chip is placed in vinegar, bubbles of gas form. The mass of the dry marble chip can be found before and after the experiment.

Question 2. Acid Rain

A marble chip has a mass of 2.0 grams before being immersed in vinegar overnight. The chip is removed and dried the next day. What will the mass of the dried marble chip be?

●	Less than 2.0 grams
B	Exactly 2.0 grams
C	Between 2.0 and 2.4 grams
D	More than 2.4 grams

Question 3: Acid Rain

Students who did this experiment also placed marble chips in pure (distilled) water overnight. Explain why the students include this step in their experiment.

To provide a control. Maybe the liquid is the problem with marble being eaten away. (full credit)

To see the difference between acidic and non-acidic water. (partial credit)

The Grand Canyon

The Grand Canyon is located in a desert in the USA. It is a very large and deep canyon containing many layers of rock. Sometime in the past, movements in the Earth's crust lifted these layers up. The Grand Canyon is now 1.6 km deep in parts. The Colorado River runs through the bottom of the canyon.

See the picture below of the Grand Canyon taken from its south rim. Several different layers of rock can be seen in the walls of the canyon.

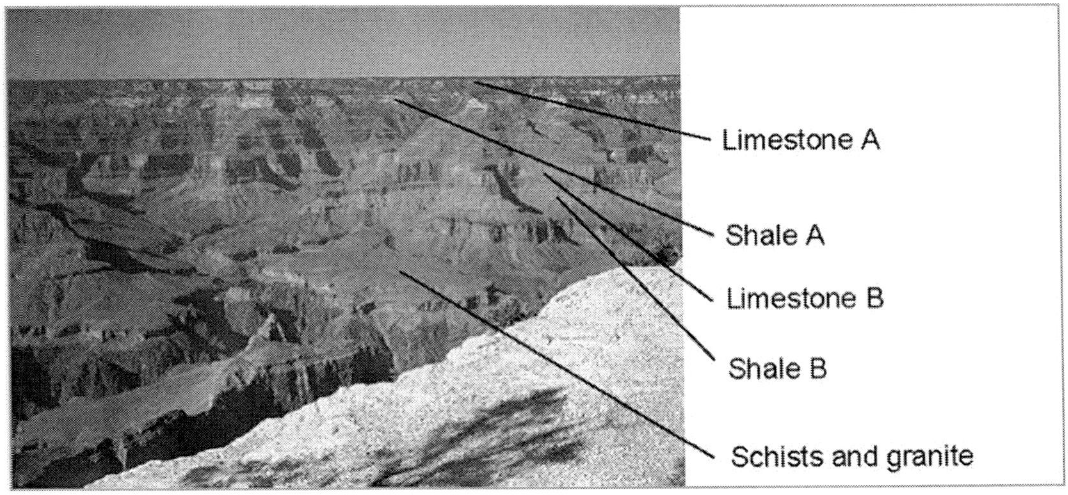

Question 1: The Grand Canyon

About five million people visit the Grand Canyon national park every year. There is concern about the damage that is being caused to the park by so many visitors.

Can the following questions be answered by scientific investigation? Circle "Yes" or "No" for each question.

Can this question be answered by scientific investigation?	Yes or No?
How much erosion is caused by use of the walking tracks?	•/No
Is the park area as beautiful as it was 100 years ago?	Yes/•

Question 2. The Grand Canyon

The temperature in the Grand Canyon ranges from below 0° C to over 40° C. Although it is a desert area, cracks in the rocks sometimes contain water. How do these temperature changes and the water in rock cracks help to speed up the breakdown of rocks?

A Freezing water dissolves warm rocks.
B Water cements rocks together.
C Ice smoothes the surface of rocks.
• Freezing water expands in the rock cracks.

Question 3. The Grand Canyon

There are many fossils of marine animals, such as clams, fish and corals, in the Limestone A layer of the Grand Canyon. What happened millions of years ago that explains why such fossils are found there?

A In ancient times, people brought seafood to the area from the ocean.
B Oceans were once much rougher and sea life washed inland on giant waves.
• An ocean covered this area at that time and then receded later.
D Some sea animals once lived on land before migrating to the sea.

Sunscreens

Mimi and Dean wondered which sunscreen product provides the best protection for their skin. Sunscreen products have a *Sun Protection Factor (SPF)* that shows how well each product absorbs the ultraviolet radiation component of sunlight. A high SPF sunscreen protects skin for longer than a low SPF sunscreen.

Mimi thought of a way to compare some different sunscreen products. She and Dean collected the following:

- two sheets of clear plastic that do not absorb sunlight;
- one sheet of light-sensitive paper;

- mineral oil (M) and a cream containing zinc oxide (ZnO); and
- four different sunscreens that they called S1, S2, S3, and S4.

Mimi and Dean included mineral oil because it lets most of the sunlight through, and zinc oxide because it almost completely blocks sunlight.

Dean placed a drop of each substance inside a circle marked on one sheet of plastic, then put the second plastic sheet over the top. He placed a large book on top of both sheets and pressed down.

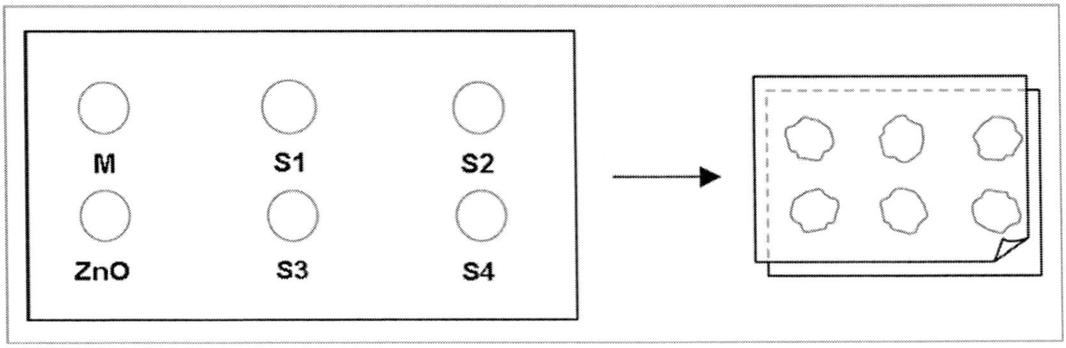

Mimi then put the plastic sheets on top of the sheet of light-sensitive paper. Light-sensitive paper changes from dark gray to white (or very light gray), depending on how long it is exposed to sunlight. Finally, Dean placed the sheets in a sunny place.

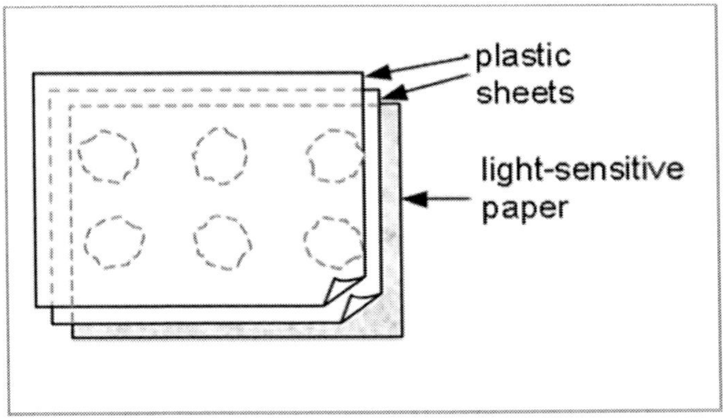

Question 1. Sunscreens

Which one of these statements is a scientific description of the role of the mineral oil and the zinc oxide in comparing the effectiveness of the sunscreens?

A Mineral oil and zinc oxide are both factors being tested.

B Mineral oil is a factor being tested and zinc oxide is a reference substance.

C Mineral oil is a reference substance and zinc oxide is a factor being tested.

● Mineral oil and zinc oxide are both reference substances.

Question 2. Sunscreens

Which one of these questions were Mimi and Dean trying to answer?

- • How does the protection for each sunscreen compare with the others?
- B How do sunscreens protect your skin from ultraviolet radiation?
- C Is there any sunscreen that gives less protection than mineral oil?
- D Is there any sunscreen that gives more protection than zinc oxide?

Question 3. Sunscreens

Why was the second sheet of plastic pressed down?

- A To stop the drops from drying out.
- B To spread the drops out as far as possible.
- C To keep the drops inside the marked circles.
- • To make the drops the same thickness.

Question 4. Sunscreens

The light-sensitive paper is a dark gray and fades to a lighter gray when it is exposed to some sunlight, and to white when exposed to a lot of sunlight.

Which one of these diagrams shows a pattern that might occur? Explain why you chose it.

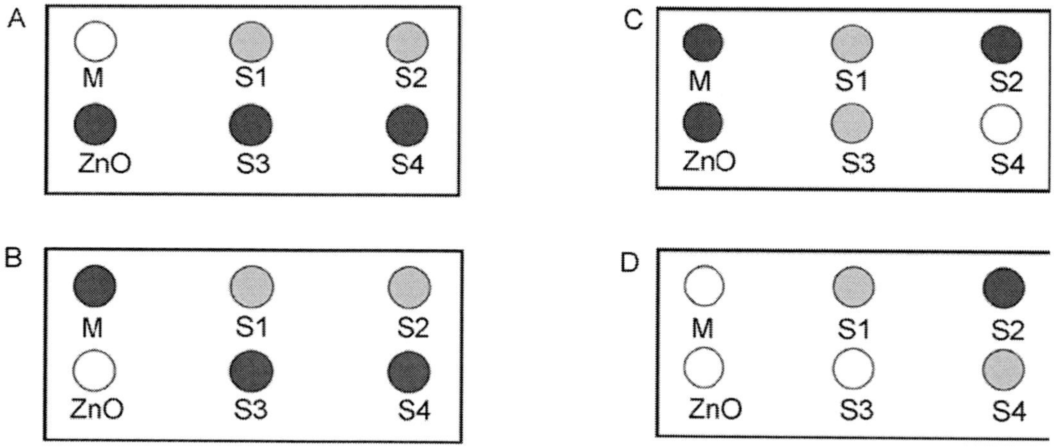

Answer: A.

Explanation: Mineral oil lets in a lot of sunlight, so that spot on the paper should be the lightest. Zinc oxide almost completely blocks sunlight, so that should be the darkest spot on the paper. (full credit)

Answer: A.

Explanation: Because ZnO blocks the light and M absorbs it. (partial credit)

APPENDIX B: TECHNICAL NOTES

The Program for International Student Assessment (PISA) is a system of international assessments that measures 15-year-olds' performance in reading literacy, mathematics literacy, and science literacy. PISA was first implemented in 2000 and is carried out every 3 years by the Organization for Economic Cooperation and Development (OECD). In this third cycle, PISA 2006, science literacy was the major focus. This appendix describes features of the PISA 2006 survey methodology, including sample design, test design, scoring, data reliability, and analysis variables. For further details about the assessment and any of the topics discussed here, see the OECD's PISA 2006 Technical Report (Adams in press) and the PISA 2003 Technical Report (Adams 2004).

International Requirements for Sampling, Data Collection, and Response Rates

To provide valid estimates of student achievement and characteristics, the sample of PISA students had to be selected in a way that represented the full population of 15-year-old students in each jurisdiction. The international desired population in each jurisdiction consisted of 15-year-olds attending both publicly and privately controlled schools in grade 7 and higher. A minimum of 4,500 students from a minimum of 150 schools was required. Within schools, a sample of 35 students was to be selected in an equal probability sample unless fewer than 35 students age 15 were available (in which case all students were selected). International standards required that students in the sample be 15 years and 3 months to 16 years and 2 months at the beginning of the testing period. The testing period suggested by the OECD was between March 1, 2006, and August 31, 2006, and was required not to exceed 42 days.[12] Each jurisdiction collected its own data, following international guidelines and specifications.

The school response rate target was 85 percent for all jurisdictions. A minimum of 65 percent of schools from the original sample of schools were required to participate for a jurisdiction's data to be included in the international database. Jurisdictions were allowed to use replacement schools (selected during the sampling process) to increase the response rate once the 65 percent benchmark had been reached.

PISA 2006 also required a minimum participation rate of 80 percent of sampled students from schools within each jurisdiction. A student was considered to be a participant if he or she participated in the first testing session or a follow-up or makeup testing session. Data from jurisdictions not meeting this requirement could be excluded from international reports.

[12] The United States, the United Kingdom (except Scotland), and Bulgaria were given permission to move the testing dates to the fall in an effort to improve response rates. The range of eligible birthdates was adjusted so that the mean age remained the same. In 2003, the United States conducted PISA in the spring and fall and found no significant difference in student performance between the two time points.

Exclusion guidelines allowed for 5 percent at the school level for approved reasons (for example, remote regions or very small schools) and 2 percent for special education schools. Overall estimated student exclusions were to be under 5 percent. PISA's intent was to be as inclusive as possible. A special 1-hour test booklet was developed for use in special education classrooms, and jurisdictions could choose whether or not to use the booklet. The United States chose not to use this special test booklet.

Schools used the following international guidelines on possible student exclusions:

- **Functionally disabled students.** These were students with a moderate to severe permanent physical disability such that they cannot perform in the PISA testing environment
- **Intellectually disabled students.** These were students with a mental or emotional disability and who have been tested as cognitively delayed or who are considered in the professional opinion of qualified staff to be cognitively delayed such that they cannot perform in the PISA testing situation.
- **Students with insufficient language experience.** These were students who meet the three criteria of not being native speakers in the assessment language, having limited proficiency in the assessment language, and receiving less than 1 year of instruction in the assessment language.

Quality monitors from the PISA Consortium visited a sample of schools in every jurisdiction to ensure that testing procedures were carried out in a consistent manner.

Sampling, Data Collection, and Response Rates in the United States

The PISA 2006 school sample was drawn for the United States in June 2005 by the international PISA Consortium. Unlike the 2000 PISA sample, which had a three-stage design, the U.S. sample for 2006 followed the model used in 2003, which was a two- stage sampling process with the first stage a sample of schools and the second stage a sample of students within schools. For PISA 2000, the U.S. school sample had the selection of a sample of geographic Primary Sampling Units (PSUs) as the first stage of selection. The sample was not clustered at the geographic level for PISA 2006 or PISA 2003. This change was made in an effort to reduce the design effects observed in the 2000 data and to spread the respondent burden across school districts as much as possible. The sample design for PISA 2006 was a stratified systematic sample, with sampling probabilities proportional to measures of school size. The PISA sample was stratified into two explicit groups: large schools and small schools. The frame was implicitly stratified (i.e., sorted for sampling) by five categorical stratification variables: grade span of the school (five levels), control of school (public or private), region of the country (Northeast, Central, West, Southeast)[13], type of location

[13] The Northeast region consists of Connecticut, Delaware, the District of Columbia, Maine, Maryland, Massachusetts, New Hampshire, New Jersey, New York, Pennsylvania, Rhode Island, and Vermont. The Central region consists of Illinois, Indiana, Iowa, Kansas, Michigan, Minnesota, Missouri, Nebraska, North Dakota, Ohio, Wisconsin, and South Dakota. The West region consists of Alaska, Arizona, California, Colorado, Hawaii, Idaho, Montana, Nevada, New Mexico, Oklahoma, Oregon, Texas, Utah, Washington, and Wyoming. The Southeast

relative to populous areas (eight levels), and proportion of non-white students (above or below 15 percent). The last variable used for sorting within the implicit stratification was by estimated enrollment of 15-year-olds based on grade enrollments.

Following the PISA guidelines at the same time as the PISA sample was selected, replacement schools were identified by assigning the two schools neighboring the sampled school in the frame as replacements. There were several constraints on the assignment of substitutes. One sampled school was not allowed to substitute for another, and a given school could not be assigned to substitute for more than one sampled school. Furthermore, substitutes were required to be in the same implicit stratum as the sampled school. If the sampled school was the first or last school in the stratum, then the second school following or preceding the sampled school was identified as the substitute. One was designated a first replacement and the other a second replacement. If an original school refused to participate, the first replacement was then contacted. If that school also refused to participate, the second school was then contacted.

The U.S. PISA 2006 school sample consisted of 236 schools. This number was increased from the international minimum requirement of 150 to offset school nonresponse and reduce design effects. The schools were selected with probability proportionate to the school's estimated enrollment of 15-yearolds from the school frame with 2003–04 school year data. The data for public schools were from the 2003–04 Common Core of Data (CCD), and the data for private schools were from the 2003–04 Private School Universe Survey (PSS). Any school containing at least one 7th- through 12th-grade class as of school year 2003–04 was included in the school sampling frame. Participating schools provided lists of 15-year-old students typically in August or September 2006, and a sample of 42 students was selected within each school in an equal probability sample. The overall sample design for the United States was intended to approximate a self-weighting sample of students as much as possible, with each 15-year-old student having an equal probability of being selected.

In the United States, for a variety of reasons reported by school administrators (such as increased testing requirements at the national, state, and local levels; concerns about the timing of the PISA assessment; and loss of learning time), many schools in the original sample declined to participate. The United States has had difficulty meeting the minimum response rate standards in prior years and, in 2003, opened a second data collection period in the fall of 2003 with the agreement of the PISA Consortium. A bias analysis conducted in 2003 found no statistically significant session effects between the spring and fall assessments. To improve response rates and better accommodate school schedules, the PISA 2006 data collection was scheduled from September to November 2006 with the agreement of the PISA Consortium. After experiencing similar difficulties in 2003, the United Kingdom (except Scotland) and Bulgaria also opted for a fall data collection period for PISA 2006.

Of the 236 original sampled schools, 209 were eligible (18 schools did not have any 15-year-olds enrolled, 5 had closed, and 4 were alternative schools for behavioral issues where students returned to a base school after a short period of time), and 145 agreed to participate. The weighted school response rate before replacement was 69 percent, placing the United States in the "intermediate" response rate category. The weighted school response rate before replacement is given by the formula

region consists of Alabama, Arkansas, Florida, Georgia, Kentucky, Louisiana, Mississippi, North Carolina, South Carolina, Tennessee, Virginia, and West Virginia.

$$\text{Weighted school response rate before replacement} = \frac{\sum_{i \in Y} WiEi}{\sum_{i \in (Y \cup N)} WiEi}$$

where *Y* denotes the set of responding original sample schools with age-eligible students; *N* denotes the set of eligible nonresponding original sample schools; *Wi* denotes the base weight for school *i*; *Wi* = 1/*Pi*, where *Pi* denotes the school selection probability for school *i*; and *Ei* denotes the enrollment size of age-eligible students, as indicated in the sampling frame.

In addition to the 145 participating original schools, 21 replacement schools also participated for a total of 166 participating schools, or a 79 percent overall response rate.[14] The participation of the additional schools did not change the classification of the United States in the intermediate response rate category.

A total of 6,796 students were sampled for the assessment. Of these students, 37 were deemed ineligible because of their enrolled grades or birthdays and 326 were deemed ineligible because they had left the school. These students were removed from the sample. Of the eligible 6,433 sampled students, an additional 254 were excluded using the decision criteria described earlier, for a weighted exclusion rate of 3.8 percent at the student level. Combined with the 0.5 percent of students excluded at the school level, before sampling, the overall exclusion rate for the United States was 4.3 percent.

Of the 6,179 remaining sampled students, a total of 5,611 participated in the assessment in the United States. An overall weighted student response of 91 percent was achieved.

A bias analysis was conducted in the United States to address potential problems in the data owing to school nonresponse (Krotki and Bland 2008). To compare PISA respondents and nonrespondents, it was necessary to match the sample of schools back to the sample frame to detect as many characteristics as possible that might provide information about the presence of nonresponse bias. Comparing frame characteristics for respondents and nonrespondents is not always a good measure of nonresponse bias if the characteristics are unrelated or weakly related to more substantive items in the survey; however, this was the only approach available given that no comparable school- or student-level achievement data were available. Frame characteristics were taken from the 2003–04 CCD for public schools and from the 2003–04 PSS for private schools. For categorical variables, response rates by characteristics were calculated. The hypothesis of independence between the characteristics and response status was tested using a Rao-Scott modified chi-square statistic. For continuous variables, summary means were calculated.

The 95 percent confidence interval for the difference between the mean for respondents and the overall mean was tested to see whether or not it included zero. In addition to these tests, logistic regression models were employed to identify whether any of the frame characteristics were significant in predicting response status. All analyses were performed using SUDAAN, a statistical software package. The school base weights used in these

---

[14] Response rates reported here are based on the formula used in the international report and are not consistent with NCES standards. A more conservative way to calculate the response rate would be to include replacement schools that participated in the denominator as well as the numerator, and to add replacement schools that were hard refusals to the denominator. This results in a response rate of 67.5 percent.

analyses did not include a nonresponse adjustment factor. The base weight for each original school was the reciprocal of its selection probability. The base weight for each replacement school was set equal to the base weight of the original school it replaced.

Characteristics available for public and private schools included public/private affiliation, community type, region, number of age-eligible students enrolled, total number of students, and percentage of various racial/ethnic groups (Asian or Pacific Islander, non- Hispanic; black, non-Hispanic; Hispanic; American Indian or Alaska Native, non-Hispanic; and white, non-Hispanic). The percentage of students eligible for free or reduced-price lunch was available for public schools only. For the original sample of schools, only one variable, community type (urban, suburban, or rural), showed a relationship to response status in tests of independence; school location in an urban fringe area or large town was associated with nonresponse. Using the same analytic procedure for the final sample (including replacement schools), tests of independence again showed that responding schools were less likely to be located in urban fringe areas or large towns. This same variable was found to be significant in the logistic regression model predicting response.

The international consortium adjusted the school base weights for nonresponse, as discussed in the section on weighting. Three variables were used that had been identified as stratification variables at the time of sampling: school control (public/private), census region, and community type (urban, suburban, rural). Because the nonresponse adjustments were done by the international consortium, the nonresponse bias analysis of the U.S. data was not used to inform the nonresponse weight adjustments. Thus, there was not an explicit nonresponse adjustment for this identified source of bias.

Test Development

The development of the PISA 2006 assessment instruments was an interactive process among the PISA Consortium, various expert committees, and OECD members. The assessment was developed by international experts and PISA Consortium test developers, and items were reviewed by representatives of each jurisdiction for possible bias and relevance to PISA's goals. The intention was to refl ect the national, cultural, and linguistic variety among OECD jurisdictions. The assessment included items submitted by participating jurisdictions as well as items that were developed by the Consortium's test developers.

The final assessment consisted of 140 science items, 48 mathematics items, and 28 reading items allocated to 13 test booklets. Each booklet was made up of 4 test clusters. Altogether there were 7 science clusters (S1–S7), 4 mathematics clusters (M1–M4), and 2 reading clusters (R1–R2). The clusters were allocated in a rotated design to the 13 booklets. The average number of items per cluster was 20 items for science, 12 items for mathematics, and 14 items for reading. Each cluster was designed to average 30 minutes of test material. Each student took one booklet, with about 2 hours worth of testing material. Approximately one-third of the science literacy items were multiple choice, one-third were closed or short response types (for which students wrote an answer that was simply either correct or incorrect), and about one-third were open constructed responses (for which students wrote answers that were graded by trained scorers using an international scoring guide). In PISA 2006, every student answered science items. Mathematics and reading items were spread throughout other booklets. The United States did not use the optional 1-hour test booklet that included lower difficulty items designed for use in special education classrooms. This booklet was used by seven jurisdictions: Austria, Belgium, the Czech Republic, Germany, the Netherlands, Slovakia, and Slovenia. For more information on assessment design, see the OECD's *PISA 2006 Technical Report* (Adams in press).

In addition to the cognitive assessment, students also received a 30-minute questionnaire designed to provide information about their backgrounds, attitudes, and experiences in school. Principals in schools where PISA was administered also received a 20- to 30- minute questionnaire about their schools. Results from the school survey are not discussed in this report but are available in *PISA 2006: Science Competencies for Tomorrow's World* (Vols. 1 and 2) (OECD, 2007a, 2007b).

Translation

Source versions of all instruments (assessment booklets, questionnaires, and manuals) were prepared in English and French and translated into the primary language or languages of instruction in each jurisdiction. PISA recommended that jurisdictions prepare and consolidate independent translations from both source versions and provided precise translation guidelines that included a description of the features each item was measuring and statistical analysis from the field trial. In cases for which one source language was used, independent translations were required and discrepancies reconciled. In addition, it was sometimes necessary to adapt the instrument for cultural purposes, even in nations such as the United States that use English as the primary language of instruction. For example, words such as "lift" might be adapted to "elevator" for the United States. The PISA Consortium verified the national translation and adaptation of all instruments. Electronic copies of printed materials were sent to the PISA Consortium for a final visual check prior to data collection.

Test Printing

An error was made in printing the final test booklets in the United States and the pagination of the booklets was consistently off by one page. The international consortium intended for the first page to be printed on the inside of the back cover; in the United States it was printed on the typical first page of plain white paper. As a result, some of the instructions in the reading section were incorrect. In some passages, students were incorrectly instructed to refer to the passage on the "opposite page" when the passage now appeared on the previous page. Because of the small number of items in the reading section, it was not possible to recalibrate the score to exclude the affected items. No incorrect page references appeared in the mathematics or science sections of the assessments. However, in some instances math and science items could be more difficult because the question required information provided previously that now required the student to turn back a page. In a few instances, items could be somewhat easier because of the pagination. ACER examined the potential impact of this on the math and science scales and estimated the scores would change by one point if the items that may have been affected by pagination were removed. Because one point is within the equating error of the scale, the original scales were retained using the results from all mathematics and science items.

Test Administration and Quality Assurance

PISA 2006 emphasized the use of standardized procedures in all jurisdictions. Each jurisdiction collected its own data, based on comprehensive manuals and training sessions provided by the PISA Consortium to explain the survey's implementation, including precise instructions for the work of school coordinators and scripts for test administrators to use in testing sessions. Test administration in the United States was carried out by professional staff trained according to the international guidelines. School staff were asked only to assist with listing students, identifying space for testing in the school, and specifying any parental consent procedures needed for sampled students. Students were allowed to use calculators, and U.S. students were provided calculators; however, no information on the availability of calculators was collected internationally.

At some schools, the PISA test was administered to students outside of normal school hours to address schools' concerns about the potential negative effect on students of the loss of instructional time. Tests were administered during normal school hours at 88 schools (53 percent), after normal school hours at 4 schools (2 percent), and on Saturday mornings at 74 schools (45 percent).

No differences were found between the schools that administered the test during out-of-school hours and the schools that opted for traditional in-school testing. Tests for differences by a variety of school characteristics (school control, locale, region, school size, school racial composition, and percentage of students receiving free or reduced-price lunch) demonstrated no significant results. Tests for differences in student test scores were implemented at both the school and student levels, and no measurable differences were found between the two groups of schools. Finally, a regression analysis of test scores as a function of selected school characteristics found no significant effect of the type of administration on the final test scores (Krotki and Bland 2008).

Members of the PISA Consortium visited all national centers to review data collection procedures, and members of the PISA Consortium also visited a randomly selected subsample of approximately 10 percent of the schools to ensure that procedures were being carried out in accordance with international guidelines. For a detailed description of the quality assurance procedures, see the OECD's *PISA 2006 Technical Report* (Adams in press).

Scoring

At least one-third of the PISA assessment was devoted to items requiring constructed responses. The process of scoring these items was an important step in ensuring the quality and comparability of the PISA data. Detailed guidelines were developed for the scoring guides themselves, training materials to recruit scorers, and workshop materials used for the training of national scorers. Prior to the national training, the PISA Consortium organized training sessions to present the material and train the scoring coordinators from the participating jurisdictions, who trained the national scorers.

For each test item, the scoring guide described the intent of the question and how to score the students' responses to each item. This description included the credit labels—full credit, partial credit, or no credit— attached to the possible categories of response. In addition, the scoring guides included real examples of students' responses accompanied by a rationale for their classification for purposes of clarity and illustration.

To examine the consistency of this marking process in more detail within each jurisdiction and to estimate the magnitude of the variance components associated with the use of scorers, the PISA Consortium conducted an interscorer reliability study on a subsample of assessment booklets. Homogeneity analysis was applied to the national sets of multiple scoring and compared with the results of the field trial. A full description of this process and the results can be found in the OECD's *PISA 2006 Technical Report* (Adams in press).

Data Entry and Cleaning

Data entry was the responsibility of the national project manager from each nation. The data collected for PISA 2006 were entered into data files with a common international format, as specified in the *PISA 2006 Main Study Management Manual, Version 3* (Australian Council for Educational Research [ACER] 2006). Data entry was completed using specialized software that allowed data to be merged into Keyquest, a common data processing software application developed by the ACER for use by participating nations. The software facilitated the checking and correction of data by providing various data consistency checks. The data were then sent to ACER for cleaning. ACER's role at this point was to check that the international data structure was followed, check the identification system within and between files, correct single case problems manually, and apply standard cleaning procedures to questionnaire files. Results of the data cleaning process were documented and shared with the national project managers and included specific questions when required. The national project manager then provided ACER with revisions to coding or solutions for anomalies. ACER then compiled background univariate statistics and preliminary classical and Rasch Item Analysis. Detailed information on the entire data entry and cleaning process can be found in the OECD's *PISA 2006 Technical Report* (Adams in press).

Weighting

The use of sampling weights is necessary for the computation of statistically sound, nationally representative estimates. Adjusted survey weights adjust for the probabilities of selection for individual schools and students, for school or student nonresponse, or for errors in estimating the size of the school or the number of 15-year-olds in the school at the time of sampling. Survey weighting for all jurisdictions participating in PISA 2006 was carried out by Westat, as part of the PISA Consortium.

The internationally defined weighting specifications for PISA 2006 included two base weights and five adjustments. The school base weight was defined as the reciprocal of the school's probability of selection. (For replacement schools, the school base weight was set equal to the original school it replaced.) The student base weight was given as the reciprocal of the probability of selection for each selected student from within a school.

The product of these base weights was then adjusted for school and student nonresponse. The school nonresponse adjustment was done individually for each jurisdiction using the implicit and explicit strata defined as part of the sample design. In the case of the United

States, three variables were used: school control, census region, and community type. The student nonresponse adjustment was done within cells based first on their final school nonresponse cell and their explicit stratum, and within that, grade and gender were used as possible. Grade and gender were collected for students in all jurisdictions on the student tracking form. Trimming factors at the school and student levels were also used (one school weight was trimmed for the United States data; no student weights were trimmed). All PISA analyses were conducted using these adjusted sampling weights. For more information on the nonresponse adjustments and trimming factors, see the OECD's *PISA 2006 Technical Report* (Adams in press).

Scaling of Student Test Data

Thirteen versions of the PISA test booklet were created, each containing a slightly different subset of items. The fact that each student completed only a subset of items means that classic test scores, such as the percent correct, are not accurate measures of student performance. Instead, scaling techniques were used to establish a common scale for all students. For PISA 2006, item response theory (IRT) was used to estimate average scores for science, mathematics, and reading literacy for each jurisdiction.

IRT identifies patterns of response and uses statistical models to predict the probability of answering an item correctly as a function of the students' proficiency in answering other questions. PISA 2006 used a mixed coefficients multinomial logit IRT model. This model is similar in principle to the more familiar two-parameter IRT model. With this method, the performance of a sample of students in a subject area or sub-area can be summarized on a simple scale or series of scales, even when students are administered different items.

Scores for students are estimated as plausible values because each student completed only a subset of items. Five plausible values were estimated for each student for each scale. These values represent the distribution of potential scores for all students in the population with similar characteristics and identical patterns of item response. Statistics describing performance on the PISA science and mathematics scales are based on plausible values.[15]

Proficiency Levels

In addition to a range of scale scores as the basic form of measurement, PISA also describes student proficiency in science literacy in terms of six described levels. Increasing levels represent the knowledge, skills, and capabilities needed to perform tasks of increasing complexity. As a result, the findings are reported in terms of percentages of the student population at each of the predefined levels.

[15] For theoretical and empirical justification of the procedures employed, see Mislevy (1988).

Exhibit B-1. Description of general competencies and examples of tasks students should be able to do, by science literacy subscale and proficiency level: 2006

	Task descriptions		
ProÞciency level	Identifying scientiÞc issues	Explaining phenomena scientiÞcally	Using scientiÞc evidence
Level 1	Students at this level should be able to suggest appropriate sources of information on scientiÞc topics. They should be able to identify a quantity that is undergoing variation in an experiment. In speciÞc contexts they should be able to recognize whether that variable can be measured using familiar measuring tools or not.	Students at this level should be able to recognize simple cause-and-effect relationships given relevant cues. The knowledge drawn upon is a singular scientiÞc fact that is drawn from experience or has widespread popular currency.	In response to a question, students at this level should be able to extract information from a fact sheet or diagram pertinent to a common context. They should be able to extract information from bar graphs where the requirement is simple comparisons of bar heights. In common, experienced contexts students at this level should be able to attribute an effect to a cause.
Level 2	Students at this level should be able to determine if scientiÞc measurement can be applied to a given variable in an investigation. They should be able to recognize the variable being manipulated (changed) by the investigator. Students should be able to appreciate the relationship between a simple model and the phenomenon it is modeling. In researching topics students should be able to select appropriate key words for a search.	Students at this level should be able to recall an appropriate, tangible, scientiÞc fact applicable in a simple and straightforward context and should be able to use it to explain or predict an outcome.	Students at this level should be able to recognize the general features of a graph if they are given appropriate cues and can point to an obvious feature in a graph or simple table in support of a given statement. They should be able to recognize if a set of given characteristics applies to the function of everyday artifacts in making choices about their use.
Level 3	Students at this level should be able to make judgments about whether an issue is open to scientiÞc measurement and, consequently, to scientiÞc investigation. Given a description of an investigation, they should be able to identify the change and measured variables.	Students at this level should be able to apply one or more concrete or tangible scientiÞc ideas/concepts in the development of an explanation of a phenomenon. This is enhanced when there are speciÞc cues given or options available from which to choose. When developing an explanation, cause-and-effect relationships are recognized and simple, explicit scientiÞc models may be drawn upon.	Students at this level should be able to select a piece of relevant information from data in answering a question or in providing support for or against a given conclusion. They should be able to draw a conclusion from an uncomplicated or simple pattern in a dataset. Students should be able to also determine, in simple cases, if enough information is present to support a given conclusion.
Level 4	Students at this level should be able to identify the change and measured variables in an investigation and at least one variable that is being controlled. They should be able to suggest appropriate ways of controlling that variable. The question being investigated in straightforward investigations can be articulated.	Students at this level should have an understanding of scientiÞc ideas, including scientiÞc models, with a signiÞcant level of abstraction. They should be able to apply a general, scientiÞc concept containing such ideas in the development of an explanation of a phenomenon.	Students at this level should be able to interpret a dataset expressed in a number of formats, such as tabular, graphic, and diagrammatic, by summarizing the data and explaining relevant patterns. They should be able to use the data to draw relevant conclusions. Students should also be able to determine whether the data support assertions about a phenomenon.
Level 5	Students at this level understand the essential elements of a scientiÞc investigation and thus should be able to determine if scientiÞc methods can be applied in a variety of quite complex, and often abstract contexts. Alternatively, by analyzing a given experiment they should be able to identify the question being investigated and explain how the methodology relates to that question.	Students at this level should be able to draw on knowledge of two or three scientiÞc concepts and identify the relationship between them in developing an explanation of a contextual phenomenon.	Students at this level should be able to interpret data from related datasets presented in various formats. They should be able to identify and explain differences and similarities in the datasets and draw conclusions based on the combined evidence presented in those datasets.
Level 6	Students at this level should demonstrate an ability to understand and articulate the complex modeling inherent in the design of an investigation.	Students at this level should be able to draw on a range of abstract scientiÞc knowledge and concepts and the relationships between these in developing explanations of processes within systems.	Students at this level should demonstrate an ability to compare and differentiate among competing explanations by examining supporting evidence. They should be able to formulate arguments by synthesizing evidence from multiple sources.

NOTE: To reach a particular proÞciency level, a student must correctly answer a majority of items at that level. Students were classiÞed into science literacy levels according to their scores. Exact cut point scores are as follows: below level 1 (a score less than or equal to 334.94); level 1 (a score greater than 334.94 and less than or equal to 409.54); level 2 (a score greater than 409.54 and less than or equal to 484.14); level 3 (a score greater than 484.14 and less than or equal to 558.73); level 4 (a score greater than 558.73 and less than or equal to 633.33); level 5 (a score greater than 633.33 and less than or equal to 707.93); and level 6 (a score greater than 707.93).
SOURCE: Organization for Economic Cooperation and Development (OECD). (2006). *Assessing Scienti c, Reading and Mathematical Literacy: A Framework for PISA 2006.* Paris: Author; Organization for Economic Cooperation and Development (OECD), Program for International Student Assessment (PISA), 2006.

Each of the four science literacy scales—the combined scale and the three subscales—is divided into six levels. Descriptions were developed to characterize typical student performance at each level. A seventh level (below level 1) was established to include students whose abilities could not be accurately described based on their responses. Exhibit 1 in the body of the report summarizes the knowledge and skills that students need to demonstrate to be classified into one of the six levels on the combined science literacy scale. Similarly, exhibit B-1 in this appendix presents the proficiency descriptions for each of the six levels on the science subscales. Exact cut scores for the levels are as follows: below level 1 (a score less than or equal to 334.94); level 1 (a score greater than 334.94 and less than or equal to 409.54); level 2 (a score greater than 409.54 and less than or equal to 484.14); level 3 (a score greater than 484.14 and less than or equal to 558.73); level 4 (a score greater than 558.73 and less than or equal to 633.33); level 5 (a score greater than 633.33 and less than or equal to 707.93); and level 6 (a score greater than 707.93).

To determine the performance levels and cut scores on the literacy scales, IRT techniques were used. With IRT techniques, it is possible to simultaneously estimate the ability of all students taking the PISA assessment, as well as the difficulty of all PISA items. Then estimates of student ability and item difficulty can be mapped on a single continuum. The relative ability of students taking a particular test can be estimated by considering the percentage of test items they get correct. The relative difficulty of items in a test can be estimated by considering the percentage of students getting each item correct. In PISA, all students within a level are expected to answer at least half of the items from that level correctly. Students at the bottom of a level are able to provide the correct answers to about 52 percent of all items from that level, have a 62 percent chance of success on the easiest items from that level, and have a 42 percent chance of success on the hardest items from that level. Students in the middle of a level have a 62 percent chance of correctly answering items of average difficulty for that level (an overall response probability of 62 percent). Students at the top of a level are able to provide the correct answers to about 70 percent of all items from that level, have a 78 percent chance of success on the easiest items from that level, and have a 62 percent chance of success on the hardest items from that level. Students just below the top of a level would score less than 50 percent on an assessment at the next higher level. Students at a particular level demonstrate not only the knowledge and skills associated with that level but also the proficiencies defined by lower levels. Thus, all students proficient at level 3 are also proficient at levels 1 and 2. Patterns of responses for students below level 1 suggest that these students are unable to answer at least half of the items from level 1 correctly. For details about the approach to defining and describing the PISA levels and establishing the cut scores, see the OECD's *PISA 2006 Technical Report* (Adams in press) and the *PISA 2003 Technical Report* (Adams 2004).

Data Limitations

As with any study, there are limitations to PISA 2006 that researchers should take into consideration. Estimates produced using data from PISA 2006 are subject to two types of error: nonsampling and sampling errors. Nonsampling errors can be due to errors made in the collection and processing of data. Sampling errors can occur because the data were collected from a sample rather than a complete census of the population.

Nonsampling Errors

"Nonsampling error" is a term used to describe variations in the estimates that may be caused by population coverage limitations, nonresponse bias, and measurement error, as well as data collection, processing, and reporting procedures. For example, the sampling frame was limited to regular public and private schools in the 50 states and the District of Columbia and cannot be used to represent Puerto Rico or other jurisdictions. The sources of nonsampling errors are typically problems such as unit and item nonresponse, the differences in respondents' interpretations of the meaning of survey questions, response differences related to the particular time the survey was conducted, and mistakes in data preparation. Some of these issues (particularly unit nonresponse) are discussed above in the section on U.S. sampling and data collection. Another example of nonsampling error that affected this data collection was the printing error, described earlier in the Test Printing section.

Sampling Errors

Sampling errors occur when a discrepancy between a population characteristic and the sample estimate arises because not all members of the target population are sampled for the survey. The size of the sample relative to the population and the variability of the population characteristics both infl uence the magnitude of sampling error. The particular sample of 15-year-old students from fall 2006 was just one of many possible samples that could have been selected. Therefore, estimates produced from the PISA 2006 sample may differ from estimates that would have been produced had another sample of students been selected. This type of variability is called sampling error because it arises from using a sample of 15-year-old students in 2006 rather than all 15-yearold students in that year.

One potential source of sampling error for PISA 2006 is that the weight for a replacement school was based on the weight for the school originally selected. These schools were typically very similar in size and other characteristics (the replacement schools were adjacent to the original school on the sorted list of schools), however, there could be some error associated with this method. A second potential source of sampling error could occur if the enrollment lists used for sampling were not up to date.

The standard error is a measure of the variability owing to sampling when estimating a statistic. The approach used for calculating sampling variances in PISA was the Fay method of Balanced Repeated Replication (BRR). This method of producing standard errors uses information about the sample design to produce more accurate standard errors than would be produced using simple random sample assumptions. Thus, the standard errors that are reported here can be used as a measure of the precision expected from this particular sample.

Standard errors for all of the estimates are in appendix C of this report. These standard errors can be used to produce confidence intervals. In keeping with NCES standards, 95 percent confidence intervals are used for this report. A 95 percent confidence interval is interpreted as a 95 percent chance that the true average in the population lies within the range of 1.96 times the standard error above or below the estimated score.

Missing Data

There are four kinds of missing data at the item level. "Nonresponse" data occurs when a respondent is expected to answer an item but no response is given. Responses that are "missing or invalid" occur in multiple-choice items for which an invalid response is given. The missing or invalid code is not used for open-ended questions. An item is "not applicable" when it is not possible for the respondent to answer the question. Finally, items that are "not reached" are consecutive missing values starting from the end of each test session. All four kinds of missing data are coded differently in the PISA 2006 database.

Background data were not imputed for cases with missing data, and those cases were not included in instances where they had missing data. Item response rates for variables discussed in this report were all over 85 percent. Response rates for sex were 100 percent in all participating jurisdictions and the response rate for race/ethnicity in the United States was 98 percent.

Descriptions of Background Variables

In this report, PISA 2006 results are provided for groups of students with different demographic characteristics. Definitions of subpopulations are as follows:

Sex: Results are reported separately for male students and female students.

Race/ethnicity: In the United States, students' race/ ethnicity was obtained through student responses to a two-part question in the student questionnaire. Students were asked first whether they were Hispanic or Latino and then whether they were members of the following racial groups: white (non-Hispanic), black (non-Hispanic), Asian (non-Hispanic), American Indian or Alaska Native (non-Hispanic), or Native Hawaiian/Other Pacific Islander (non-Hispanic). Multiple responses to the race classification question were allowed. Results are shown separately for white (non-Hispanic) students, black (non-Hispanic) students, Hispanic students, Asian (non-Hispanic) students, American Indian or Alaska Native (non- Hispanic) students, Native Hawaiian/Other Pacific Islander (non-Hispanic) students, and non-Hispanic students who selected more than one race. Students identifying themselves as Hispanic and one or more race were included in the Hispanic group, rather than in a racial group.

Full PISA 2006 student and school questionnaires are available at http://nces.ed.gov/surveys/pisa and http://www.pisa.oecd.org.

Confidentiality and Disclosure Limitations

The PISA 2006 data are hierarchical and include school and student data from the participating schools. Confidentiality analyses for the United States were designed to provide reasonable assurance that public-use data files issued by the PISA Consortium would not allow identification of individual U.S. schools or students when compared against other public-use data collections. Disclosure limitation included identifying and masking potential disclosure risk to PISA schools and including an additional measure of uncertainty to school and student identification through random swapping of data elements within the student and school files.

Statistical Procedures

Tests of Significance

Comparisons made in the text of this report have been tested for statistical significance. For example, in the commonly made comparison of jurisdiction averages against the average of the United States, tests of statistical significance were used to establish whether or not the observed differences from the U.S. average were statistically significant.

The estimation of the standard errors that are required in order to undertake the tests of significance is complicated by the complex sample and assessment designs, both of which generate error variance. Together they mandate a set of statistically complex procedures for estimating the correct standard errors. As a consequence, the estimated standard errors contain a sampling variance component estimated by BRR. Where the assessments are concerned, there is an additional imputation variance component arising from the assessment design. Details on the BRR procedures used can be found in the *PISA 2006 Technical Report* (Adams in press) and the *PISA 2003 Technical Report* (Adams 2004).

In almost all instances, the tests for significance used were standard *t* tests. These fell into two categories according to the nature of the comparison being made: comparisons of independent samples and comparisons of nonindependent samples. In PISA, jurisdiction samples are independent.

In simple comparisons of independent averages, such as the average score of jurisdiction 1 with that of jurisdiction 2, the following formula was used to compute the *t* statistic:

$$t = est_1 - est_2 \ / \ \text{SQRT}\ [(se_1)^2 + (se_2)^2],$$

where est_1 and est_2 are the estimates being compared (e.g., averages of jurisdiction 1 and jurisdiction 2) and se_1 and se_2 are the corresponding standard errors of these averages.

The second type of comparison used in this report occurred when comparing differences of nonsubset, nonindependent groups. When this occurs, the correlation and related covariance between the groups must be taken into account, such as when comparing the average scores of males versus females within the United States.

How are scores such as those for males and females correlated? Suppose that in the school sample, a coeducational school attended by low achievers is replaced by a coeducational school attended by high achievers. The jurisdiction mean will increase slightly, as well as the means for males and females. If such a school replacement process is continued, the average scores of males and the average scores of females will likely increase in a similar pattern. Indeed, a coeducational school attended by high- achieving males is usually also attended by high- achieving females. Therefore, the covariance between the males' scores and the females' scores is likely to be positive.

To determine whether the performance of females differs from the performance of males, the standard error of the difference that takes into account the covariance between the females' scores and the males' scores needs to be estimated. The estimation of the covariance requires the selection of several samples and then the analysis of the variation of the males' means in conjunction with the females' means. Such a procedure is, of course, unrealistic. Therefore, as for any computation of a standard error in PISA, replication methods using the supplied replicate weights were used to estimate the standard error of a difference. Use of the replicate weights implicitly incorporates the covariance between the two estimates into the estimate of the standard error of the difference.

To test such comparisons, the following formula was used to compute the t statistic:

$$t = est_{grp1} - est_{grp2} \;{}^{se}\; (est_{grp1} - est\ _{grp2}),$$

where est_{grp1} and est_{grp2} are the nonindependent group estimates being compared and $se\ (est_{grp1} - est\ _{grp2})$ is the standard error of the difference calculated using BRR to account for any covariance between the estimates for the two nonindependent groups.

Effect Size

Tests of statistical significance are, in part, infl uenced by sample sizes. To provide the reader with an increased understanding of the importance of the significant difference between student populations in the United States, effect sizes are included in the report. Effect sizes use standard deviations, rather than standard errors, and are therefore not infl uenced by the size of the student samples. Following Cohen (1988) and Rosnow and Rosenthal (1996), effect size is calculated by finding the difference between the means of two groups and dividing that result by the pooled standard deviation of the two groups:

$$d = \frac{est_{grp1} - est_{grp2}}{sd_{pooled}},$$

where est_{grp1} and est_{grp2} are the student group estimates being compared and sd_{pooled} is the pooled standard deviation of the groups being compared.

The formula for the pooled standard deviation is as follows (Rosnow and Rosenthal 1996):

$$sd_{pooled} = \frac{\sqrt{sd_1^2 + sd_2^2}}{2},$$

where sd_1 and sd_2 are the standard deviations of the groups being compared.

Table C-1. Percentage distribution of 15-year-old students, by grade level and jurisdiction: 2006

Jurisdiction	Percent	*s.e.*	Percent	*s.e.*	Percent	*s.e.*	Percent	*s.e.*	Percent	*s.e.*	Percent	*s.e.*	Percent	*s.e.*
OECD average[1]	0.9	*0.06*	5.9	*0.11*	37.5	*0.18*	48.8	*0.19*	11.5	*0.08*	0.8	*0.04*	1.7	—
OECD jurisdictions														
Australia	#	*†*	0.1	*0.03*	8.9	*0.59*	70.6	*0.93*	20.3	*0.77*	0.1	*0.05*	#	*†*
Austria	0.3	*0.11*	6.4	*0.70*	44.6	*1.13*	48.7	*1.06*	#	*†*	#	*†*	#	*†*
Belgium	0.4	*0.09*	4.4	*0.32*	31.1	*0.79*	63.2	*0.79*	1.0	*0.11*	#	*†*	#	*†*
Canada	#	*†*	1.7	*0.22*	13.3	*0.59*	83.8	*0.65*	1.2	*0.14*	#	*†*	#	*†*
Czech Republic	0.7	*0.14*	3.5	*0.39*	44.3	*1.32*	51.5	*1.49*	#	*†*	#	*†*	#	*†*
Denmark	0.2	*0.07*	12.0	*0.57*	85.3	*0.73*	1.4	*0.24*	1.1	*0.36*	#	*†*	#	*†*
Finland	0.2	*0.06*	11.7	*0.52*	88.1	*0.52*	#	*†*	#	*†*	#	*†*	#	*†*
France	#	*†*	5.2	*0.45*	34.8	*1.19*	57.5	*1.21*	2.4	*0.26*	#	*†*	#	*†*
Germany	1.5	*0.25*	11.9	*0.58*	54.5	*0.65*	28.2	*0.82*	0.3	*0.06*	#	*†*	3.6	*0.32*
Greece	0.5	*0.14*	2.1	*0.37*	5.3	*0.76*	78.7	*1.02*	13.3	*0.56*	#	*†*	#	*†*
Hungary	2.2	*0.45*	5.5	*0.62*	65.7	*0.85*	26.6	*0.54*	#	*†*	#	*†*	#	*†*
Iceland	#	*†*	#	*†*	0.2	*0.08*	99.2	*0.11*	0.6	*0.08*	#	*†*	#	*†*
Ireland	#	*†*	2.7	*0.38*	58.5	*0.79*	21.2	*1.27*	17.5	*1.08*	#	*†*	#	*†*
Italy	0.3	*0.10*	1.5	*0.43*	15.0	*0.58*	80.4	*0.69*	2.8	*0.24*	#	*†*	#	*†*
Japan	#	*†*	#	*†*	#	*†*	100.0	*0.00*	#	*†*	#	*†*	#	*†*
Korea, Republic of	#	*†*	#	*†*	2.0	*0.57*	97.3	*0.58*	0.7	*0.11*	#	*†*	#	*†*
Luxembourg	0.2	*0.07*	11.8	*0.26*	53.4	*0.43*	34.4	*0.41*	0.1	*0.04*	#	*†*	#	*†*
Mexico	2.3	*0.23*	8.1	*0.77*	33.2	*1.92*	48.5	*1.90*	5.1	*0.36*	2.0	*0.16*	0.9	*0.29*
Netherlands	0.1	*0.09*	3.7	*0.39*	44.9	*1.09*	50.7	*1.17*	0.4	*0.10*	#	*†*	#	*†*
New Zealand	#	*†*	#	*†*	#	*†*	6.2	*0.36*	89.4	*0.46*	4.4	*0.31*	#	*†*
Norway	#	*†*	#	*†*	0.5	*0.11*	99.0	*0.33*	0.5	*0.31*	#	*†*	#	*†*
Poland	0.6	*0.16*	3.8	*0.34*	95.1	*0.41*	0.6	*0.08*	#	*†*	#	*†*	#	*†*
Portugal	6.4	*0.67*	12.8	*0.68*	28.9	*1.12*	49.6	*1.53*	0.2	*0.05*	#	*†*	2.1	*0.63*
Slovak Republic	0.7	*0.21*	2.2	*0.42*	38.5	*2.09*	58.7	*2.20*	#	*†*	#	*†*	#	*†*
Spain	0.1	*0.04*	7.0	*0.47*	33.0	*0.79*	59.8	*0.87*	#	*†*	#	*†*	#	*†*
Sweden	#	*†*	1.9	*0.21*	95.9	*0.38*	2.2	*0.32*	#	*†*	#	*†*	#	*†*
Switzerland	0.8	*0.12*	16.1	*0.78*	62.6	*1.46*	20.3	*1.65*	0.3	*0.13*	#	*†*	#	*†*
Turkey	0.8	*0.32*	4.5	*0.90*	38.4	*1.73*	53.7	*1.88*	2.6	*0.27*	#	*†*	#	*†*
United Kingdom	#	*†*	#	*†*	#	*†*	0.9	*0.10*	98.4	*0.13*	0.7	*0.06*	#	*†*
United States	0.8	*0.74*	1.0	*0.87*	10.7	*0.78*	70.9	*1.42*	16.5	*0.75*	0.1	*0.05*	#	*†*
Non-OECD jurisdictions														
Argentina	3.9	*0.83*	9.4	*0.76*	17.0	*1.35*	64.4	*2.11*	3.0	*0.40*	0.6	*0.55*	1.7	*0.98*
Azerbaijan	0.5	*0.11*	5.5	*0.55*	53.5	*1.48*	39.0	*1.54*	0.6	*0.13*	0.5	*0.40*	0.5	*0.21*
Bulgaria	0.3	*0.14*	7.1	*0.96*	74.3	*1.17*	18.2	*0.90*	#	*†*	#	*†*	#	*†*
Brazil	11.6	*0.69*	22.0	*1.25*	47.8	*1.24*	18.0	*0.86*	0.6	*0.18*	#	*†*	#	*†*
Chile	1.0	*0.31*	3.3	*0.52*	18.9	*0.99*	70.8	*1.19*	6.1	*0.46*	#	*†*	#	*†*
Chinese Taipei	#	*†*	#	*†*	36.3	*1.30*	63.6	*1.32*	0.1	*0.08*	#	*†*	#	*†*
Colombia	6.4	*0.96*	12.3	*0.91*	22.2	*0.83*	37.8	*1.39*	21.4	*2.14*	#	*†*	#	*†*
Croatia	#	*†*	0.4	*0.26*	77.1	*0.48*	22.6	*0.43*	#	*†*	#	*†*	#	*†*
Estonia	3.3	*0.37*	25.6	*0.84*	69.4	*0.86*	1.8	*0.17*	#	*†*	#	*†*	#	*†*
Hong Kong-China	2.4	*0.22*	9.3	*0.54*	25.2	*0.46*	63.0	*0.93*	0.1	*0.15*	#	*†*	#	*†*
Indonesia	0.1	*0.05*	12.0	*1.68*	40.0	*2.97*	43.5	*3.76*	4.4	*0.63*	#	*†*	#	*†*
Israel	#	*†*	0.3	*0.07*	14.6	*1.05*	84.7	*1.07*	0.4	*0.10*	#	*†*	#	*†*
Jordan	0.1	*0.08*	1.3	*0.18*	8.1	*0.58*	90.5	*0.73*	#	*†*	#	*†*	#	*†*
Kyrgyz Republic	0.2	*0.10*	7.7	*0.59*	67.6	*1.22*	24.2	*1.35*	0.4	*0.13*	#	*†*	#	*†*
Latvia	2.6	*0.64*	16.4	*0.78*	77.7	*1.14*	3.0	*0.40*	#	*†*	#	*†*	0.4	*0.18*
Liechtenstein	#	*†*	16.7	*0.63*	72.0	*0.57*	11.0	*0.55*	0.3	*0.30*	#	*†*	#	*†*
Lithuania	0.9	*0.15*	12.1	*0.81*	80.0	*0.87*	6.8	*0.48*	#	*†*	#	*†*	0.2	*0.16*
Macao-China	7.7	*0.16*	20.6	*0.21*	34.7	*0.18*	36.5	*0.13*	0.6	*0.04*	#	*†*	#	*†*
Qatar	2.3	*0.10*	5.3	*0.13*	14.1	*0.13*	62.6	*0.17*	15.6	*0.15*	0.2	*0.06*	#	*†*
Republic of Montenegro	#	*†*	0.3	*0.15*	85.8	*0.22*	13.9	*0.17*	#	*†*	#	*†*	#	*†*
Republic of Serbia	0.1	*0.06*	1.8	*0.57*	96.6	*0.61*	1.6	*0.19*	#	*†*	#	*†*	#	*†*
Romania	0.7	*0.36*	13.5	*2.02*	82.9	*1.91*	2.9	*0.39*	#	*†*	#	*†*	#	*†*
Russian Federation	0.6	*0.14*	6.7	*0.89*	29.9	*1.58*	61.6	*2.00*	1.2	*0.22*	#	*†*	#	*†*
Slovenia	#	*†*	0.2	*0.11*	3.5	*0.33*	90.6	*0.35*	5.8	*0.21*	#	*†*	#	*†*
Thailand	#	*†*	1.3	*0.35*	30.5	*1.05*	65.2	*1.13*	3.0	*0.47*	#	*†*	#	*†*
Tunisia	11.4	*0.56*	16.7	*0.75*	21.1	*1.00*	46.6	*1.59*	4.3	*0.32*	#	*†*	#	*†*
Uruguay	7.5	*0.90*	9.8	*0.70*	17.3	*1.02*	58.9	*1.51*	6.6	*0.63*	#	*†*	#	*†*

— Not available.
† Not applicable.
# Rounds to zero.
[1] In computing the OECD average, the average for each column (grade in this case) is computed by averaging the estimates in the column but excluding those instances where no cases were reported (shown here as '#': rounds to zero). Therefore, the percentage distribution sums to greater than 100 (i.e., 107.1).
NOTE: The Organization for Economic Cooperation and Development (OECD) average is the average of the national averages of the OECD member jurisdictions. Because the Program for International Student Assessment (PISA) is principally an OECD study, the results for non-OECD jurisdictions are displayed separately from those of the OECD jurisdictions and are not included in the OECD average. Standard error is noted by *s.e.* Detail may not sum to totals because of rounding.
SOURCE: Organization for Economic Cooperation and Development (OECD), Program for International Student Assessment (PISA), 2006.

Table C-2. Average scores of 15-year-old students on combined science literacy scale and science literacy subscales, by jurisdiction: 2006

					Science literacy			
	Combined literacy	science scale	Identifying scientific		Explaining phenomena		Using evidence	scientific
Jurisdiction	Average	*s.e.*	Average	*s.e.*	Average	*s.e.*	Average	*s.e.*
OECD average	500	*0.5*	499	*0.5*	500	*0.5*	499	*0.6*
OECD jurisdictions								
Australia	527	*2.3*	535	*2.3*	520	*2.3*	531	*2.4*
Austria	511	*3.9*	505	*3.7*	516	*4.0*	505	*4.7*
Belgium	510	*2.5*	515	*2.7*	503	*2.5*	516	*3.0*
Canada	534	*2.0*	532	*2.3*	531	*2.1*	542	*2.2*
Czech Republic	513	*3.5*	500	*4.2*	527	*3.5*	501	*4.1*
Denmark	496	*3.1*	493	*3.0*	501	*3.3*	489	*3.6*
Finland	563	*2.0*	555	*2.3*	566	*2.0*	567	*2.3*
France	495	*3.4*	499	*3.5*	481	*3.2*	511	*3.9*
Germany	516	*3.8*	510	*3.8*	519	*3.7*	515	*4.6*
Greece	473	*3.2*	469	*3.0*	476	*3.0*	465	*4.0*
Hungary	504	*2.7*	483	*2.6*	518	*2.6*	497	*3.4*
Iceland	491	*1.6*	494	*1.7*	488	*1.5*	491	*1.7*
Ireland	508	*3.2*	516	*3.3*	505	*3.2*	506	*3.4*
Italy	475	*2.0*	474	*2.2*	480	*2.0*	467	*2.3*
Japan	531	*3.4*	522	*4.0*	527	*3.1*	544	*4.2*
Korea, Republic of	522	*3.4*	519	*3.7*	512	*3.3*	538	*3.7*
Luxembourg	486	*1.1*	483	*1.1*	483	*1.1*	492	*1.1*
Mexico	410	*2.7*	421	*2.6*	406	*2.7*	402	*3.1*
Netherlands	525	*2.7*	533	*3.3*	522	*2.7*	526	*3.3*
New Zealand	530	*2.7*	536	*2.9*	522	*2.8*	537	*3.3*
Norway	487	*3.1*	489	*3.1*	495	*3.0*	473	*3.6*
Poland	498	*2.3*	483	*2.5*	506	*2.5*	494	*2.7*
Portugal	474	*3.0*	486	*3.1*	469	*2.9*	472	*3.6*
Slovak Republic	488	*2.6*	475	*3.2*	501	*2.7*	478	*3.3*
Spain	488	*2.6*	489	*2.4*	490	*2.4*	485	*3.0*
Sweden	503	*2.4*	499	*2.6*	510	*2.9*	496	*2.6*
Switzerland	512	*3.2*	515	*3.0*	508	*3.3*	519	*3.4*
Turkey	424	*3.8*	427	*3.4*	423	*4.1*	417	*4.3*
United Kingdom	515	*2.3*	514	*2.3*	517	*2.3*	514	*2.5*
United States	489	*4.2*	492	*3.8*	486	*4.3*	489	*5.0*
Non-OECD jurisdictions								
Argentina	391	*6.1*	395	*5.7*	386	*6.0*	385	*7.0*
Azerbaijan	382	*2.8*	353	*3.1*	412	*3.0*	344	*4.0*
Brazil	390	*2.8*	398	*2.8*	390	*2.7*	378	*3.6*
Bulgaria	434	*6.1*	427	*6.3*	444	*5.8*	417	*7.5*
Chile	438	*4.3*	444	*4.1*	432	*4.1*	440	*5.1*
Chinese Taipei	532	*3.6*	509	*3.7*	545	*3.7*	532	*3.7*
Colombia	388	*3.4*	402	*3.4*	379	*3.4*	383	*3.9*
Croatia	493	*2.4*	494	*2.6*	492	*2.5*	490	*3.0*
Estonia	531	*2.5*	516	*2.6*	541	*2.6*	531	*2.7*

Table C-2. Continued

	Science literacy							
	Combined science literacy scale		Identifying scientific		Explaining phenomena		Using scientific evidence	
Jurisdiction	Average	*s.e.*	Average	*s.e.*	Average	*s.e.*	Average	*s.e.*
OECD average	500	*0.5*	499	*0.5*	500	*0.5*	499	*0.6*
Hong Kong-China	542	*2.5*	528	*3.2*	549	*2.5*	542	*2.7*
Indonesia	393	*5.7*	393	*5.6*	395	*5.1*	386	*7.3*
Israel	454	*3.7*	457	*3.9*	443	*3.6*	460	*4.7*
Jordan	422	*2.8*	409	*2.8*	438	*3.1*	405	*3.3*
Kyrgyz Republic	322	*2.9*	321	*3.2*	334	*3.1*	288	*3.8*
Latvia	490	*3.0*	489	*3.3*	486	*2.9*	491	*3.4*
Liechtenstein	522	*4.1*	522	*3.7*	516	*4.1*	535	*4.3*
Lithuania	488	*2.8*	476	*2.7*	494	*3.0*	487	*3.1*
Macao-China	511	*1.1*	490	*1.2*	520	*1.2*	512	*1.2*
Qatar	349	*0.9*	352	*0.8*	356	*1.0*	324	*1.2*
Republic of Montenegro	412	*1.1*	401	*1.2*	417	*1.1*	407	*1.3*
Republic of Serbia	436	*3.0*	431	*3.0*	441	*3.1*	425	*3.7*
Romania	418	*4.2*	409	*3.6*	426	*4.0*	407	*6.0*
Russian Federation	479	*3.7*	463	*4.2*	483	*3.4*	481	*4.2*
Slovenia	519	*1.1*	517	*1.4*	523	*1.5*	516	*1.3*
Thailand	421	*2.1*	413	*2.5*	420	*2.1*	423	*2.6*
Tunisia	386	*3.0*	384	*3.8*	383	*2.9*	382	*3.7*
Uruguay	428	*2.7*	429	*3.0*	423	*2.9*	429	*3.1*

Note: The Organization for Economic Cooperation and Development (OECD) average is the average of the national averages of the OECD member jurisdictions. Because the Program for International Student Assessment (PISA) is principally an OECD study, the results for non-OECD jurisdictions are displayed separately from those of the OECD jurisdictions and are not included in the OECD average. Because of an error in printing the test booklets, the United States mean performance may be misestimated by approximately 1 score point. The impact is below one standard error. For details see appendix B. Standard error is noted by *s.e.*

Source: Organization for Economic Cooperation and Development (OECD), Program for International Student Assessment (PISA), 2006.

Table C-3. Scores of 15-year-old students on combined science literacy scale at 10th and 90th percentiles, by jurisdiction: 2006

	Percentiles			
	10th		90th	
Jurisdiction	Score	*s.e.*	Score	*s.e.*
OECD average	375	*0.9*	622	*0.7*
OECD jurisdictions				
Australia	395	*3.4*	653	*2.9*
Austria	378	*6.2*	633	*3.6*
Belgium	374	*5.4*	634	*2.3*
Canada	410	*3.7*	651	*2.4*
Czech Republic	385	*5.2*	641	*4.3*
Denmark	373	*4.8*	615	*3.7*
Finland	453	*3.3*	673	*2.9*
France	359	*5.5*	623	*4.0*
Germany	381	*7.0*	642	*3.2*
Greece	353	*5.4*	589	*4.1*
Hungary	388	*4.2*	617	*3.1*
Iceland	364	*3.1*	614	*2.9*

Table C-3. Continued

	Percentiles			
	10th		90th	
Jurisdiction	Score	*s.e.*	Score	*s.e.*
OECD average	375	*0.9*	622	*0.7*
Ireland	385	*4.4*	630	*3.7*
Italy	351	*2.8*	598	*2.6*
Japan	396	*6.2*	654	*3.1*
Korea, Republic of	403	*5.7*	635	*4.7*
Luxembourg	358	*2.8*	609	*2.8*
Mexico	306	*4.2*	516	*3.0*
Netherlands	395	*5.4*	646	*3.4*
New Zealand	389	*4.5*	667	*3.3*
Norway	365	*5.6*	610	*3.5*
Poland	381	*2.9*	615	*3.3*
Portugal	357	*4.8*	588	*2.9*
Slovak Republic	368	*3.7*	609	*4.1*
Spain	370	*3.7*	604	*3.0*
Sweden	381	*4.0*	622	*2.6*
Switzerland	378	*4.9*	636	*3.8*
Turkey	325	*3.2*	540	*9.7*
United Kingdom	376	*4.3*	652	*2.9*
United States	349	*5.9*	628	*4.3*
Non-OECD jurisdictions				
Argentina	259	*9.0*	520	*6.5*
Azerbaijan	316	*2.4*	456	*6.4*
Brazil	281	*3.2*	510	*5.6*
Bulgaria	300	*7.1*	577	*8.2*
Chile	323	*4.1*	560	*6.5*
Chinese Taipei	402	*5.0*	651	*2.7*
Colombia	280	*4.5*	496	*4.6*
Croatia	383	*3.8*	604	*3.2*
Estonia	422	*3.8*	640	*3.3*
Hong Kong-China	418	*6.1*	655	*3.5*
Indonesia	307	*3.5*	488	*11.8*
Israel	310	*5.2*	601	*4.5*
Jordan	309	*4.0*	537	*4.5*
Kyrgyz Republic	220	*3.8*	428	*5.0*
Latvia	380	*4.2*	597	*3.5*
Liechtenstein	393	*12.8*	643	*9.4*
Lithuania	370	*3.2*	604	*4.2*
Macao-China	409	*2.5*	611	*1.8*
Qatar	253	*1.4*	462	*2.6*
Republic of Montenegro	312	*2.1*	517	*3.0*
Republic of Serbia	327	*4.0*	545	*3.8*
Romania	314	*5.0*	526	*5.7*
Russian Federation	364	*5.4*	596	*3.9*
Slovenia	391	*2.8*	647	*3.3*
Thailand	325	*3.4*	524	*3.8*
Tunisia	283	*3.4*	495	*6.0*
Uruguay	306	*4.9*	550	*3.6*

Note: The Organization for Economic Cooperation and Development (OECD) average is the average of the national averages of the OECD member jurisdictions. Because the Program for International Student Assessment (PISA) is principally an OECD study, the results for non-OECD jurisdictions are displayed separately from those of the OECD jurisdictions and are not included in the OECD average. Because of an error in printing the test booklets, the United States mean performance may be misestimated by approximately 1 score point. The impact is below one standard error. For details see appendix B. Standard error is noted by *s.e.*

Source: Organization for Economic Cooperation and Development (OECD), Program for International Student Assessment (PISA), 2006.

Table C-4. Standard deviations of the average scores of 15-year-old students on combined science literacy scale, by jurisdiction: 2006

Jurisdiction	Standard deviation	*s.e.*
OECD average	95	*0.3*
OECD jurisdictions		
Australia	100	*1.0*
Austria	98	*2.4*
Belgium	100	*2.0*
Canada	94	*1.1*
Czech Republic	98	*2.0*
Denmark	93	*1.4*
Finland	86	*1.0*
France	102	*2.1*
Germany	100	*2.0*
Greece	92	*2.0*
Hungary	88	*1.6*
Iceland	97	*1.2*
Ireland	94	*1.5*
Italy	96	*1.3*
Japan	100	*2.0*
Korea, Republic of	90	*2.4*
Luxembourg	97	*0.9*
Mexico	81	*1.5*
Netherlands	96	*1.6*
New Zealand	107	*1.4*
Norway	96	*2.0*
Poland	90	*1.1*
Portugal	89	*1.7*
Slovak Republic	93	*1.8*
Spain	91	*1.0*
Sweden	94	*1.4*
Switzerland	99	*1.7*
Turkey	83	*3.2*
United Kingdom	107	*1.5*
United States	106	*1.7*
Non-OECD jurisdictions		
Argentina	101	*2.6*
Azerbaijan	56	*1.9*
Brazil	89	*1.9*
Bulgaria	107	*3.2*
Chile	92	*1.8*
Chinese Taipei	94	*1.6*
Colombia	85	*1.8*
Croatia	86	*1.4*
Estonia	84	*1.1*

Table C-4. Continued

Jurisdiction	Standard deviation	*s.e.*
OECD average	95	*0.3*
Hong Kong-China	92	*1.9*
Indonesia	70	*3.3*
Israel	111	*2.0*
Jordan	90	*1.9*
Kyrgyz Republic	84	*2.0*
Latvia	84	*1.3*
Liechtenstein	97	*3.1*
Lithuania	90	*1.6*
Macao-China	78	*0.8*
Qatar	84	*0.8*
Republic of Montenegro	80	*0.9*
Republic of Serbia	85	*1.6*
Romania	81	*2.4*
Russian Federation	90	*1.4*
Slovenia	98	*1.0*
Thailand	77	*1.5*
Tunisia	82	*2.0*
Uruguay	94	*1.8*

Note: The Organization for Economic Cooperation and Development (OECD) average is the average of the national averages of the OECD member jurisdictions. Because the Program for International Student Assessment (PISA) is principally an OECD study, the results for non-OECD jurisdictions are displayed separately from those of the OECD jurisdictions and are not included in the OECD average. Because of an error in printing the test booklets, the United States mean performance may be misestimated by approximately 1 score point. The impact is below one standard error. For details see appendix B. Standard error is noted by *s.e.*

Source: Organization for Economic Cooperation and Development (OECD), Program for International Student Assessment (PISA), 2006.

Table C-5. Percentage distribution of 15-year-old students on combined science literacy scale, by proficiency level and jurisdiction: 2006

Table C-5. Percentage distrib ution of 15-year-old stu dents on combined science literacy scale, by pro ciency level and jurisdiction: 2006

Jurisdiction	Percent	*s.e.*	Percent	*s.e.*	Percent	*s.e.*	Percent	*s.e.*	Percent	*s.e.*	Percent	*s.e.*	Percent	*s.e.*
OECD average	5.2	*0.11*	14.1	0.15	24.0	0.17	27.4	0.17	20.3	0.16	7.7	0.10	1.3	0.04
OECD jurisdictions														
Australia	3.0	*0.25*	9.8	*0.46*	20.2	*0.63*	27.7	*0.51*	24.6	*0.53*	11.8	*0.53*	2.8	*0.26*
Austria	4.3	*0.88*	12.0	*0.98*	21.8	*1.05*	28.3	*1.05*	23.6	*1.12*	8.8	*0.69*	1.2	*0.20*
Belgium	4.8	*0.72*	12.2	*0.62*	20.8	*0.84*	27.6	*0.84*	24.5	*0.77*	9.1	*0.47*	1.0	*0.17*
Canada	2.2	*0.27*	7.8	*0.47*	19.1	*0.64*	28.8	*0.58*	27.7	*0.65*	12.0	*0.52*	2.4	*0.25*
Czech Republic	3.5	*0.57*	12.1	*0.84*	23.4	*1.17*	27.8	*1.09*	21.7	*0.92*	9.8	*0.86*	1.8	*0.32*
Denmark	4.3	*0.64*	14.1	*0.75*	26.0	*1.07*	29.3	*1.04*	19.5	*0.91*	6.1	*0.66*	0.7	*0.18*
Finland	0.5	*0.13*	3.6	*0.45*	13.6	*0.68*	29.1	*1.07*	32.2	*0.89*	17.0	*0.72*	3.9	*0.35*
France	6.6	*0.71*	14.5	*1.05*	22.8	*1.12*	27.2	*1.09*	20.9	*1.00*	7.2	*0.60*	0.8	*0.17*
Germany	4.1	*0.68*	11.3	*0.96*	21.4	*1.06*	27.9	*1.08*	23.6	*0.95*	10.0	*0.62*	1.8	*0.24*
Greece	7.2	*0.86*	16.9	*0.88*	28.9	*1.19*	29.4	*1.01*	14.2	*0.83*	3.2	*0.33*	0.2	*0.09*
Hungary	2.7	*0.33*	12.3	*0.83*	26.0	*1.15*	31.1	*1.07*	21.0	*0.87*	6.2	*0.57*	0.6	*0.16*
Iceland	5.8	*0.50*	14.7	*0.84*	25.9	*0.71*	28.3	*0.92*	19.0	*0.74*	5.6	*0.49*	0.7	*0.18*
Ireland	3.5	*0.47*	12.0	*0.82*	24.0	*0.97*	29.7	*0.98*	21.4	*0.87*	8.3	*0.62*	1.1	*0.19*
Italy	7.3	*0.46*	18.0	*0.62*	27.6	*0.78*	27.4	*0.61*	15.1	*0.58*	4.2	*0.31*	0.4	*0.09*
Japan	3.2	*0.45*	8.9	*0.73*	18.5	*0.86*	27.5	*0.85*	27.0	*1.14*	12.4	*0.63*	2.6	*0.33*
Korea, Republic of	2.5	*0.49*	8.7	*0.77*	21.2	*1.05*	31.8	*1.17*	25.5	*0.91*	9.2	*0.83*	1.1	*0.29*
Luxembourg	6.5	*0.39*	15.6	*0.65*	25.4	*0.66*	28.6	*0.93*	18.1	*0.71*	5.4	*0.34*	0.5	*0.11*
Mexico	18.2	*1.22*	32.8	*0.89*	30.8	*0.95*	14.8	*0.66*	3.2	*0.34*	0.3	*0.09*	#	†
Netherlands	2.3	*0.38*	10.7	*0.88*	21.1	*0.98*	26.9	*0.87*	25.8	*1.04*	11.5	*0.81*	1.7	*0.24*
New Zealand	4.0	*0.43*	9.7	*0.58*	19.7	*0.80*	25.1	*0.71*	23.9	*0.81*	13.6	*0.74*	4.0	*0.37*
Norway	5.9	*0.84*	15.2	*0.84*	27.3	*0.79*	28.5	*0.99*	17.1	*0.72*	5.5	*0.44*	0.6	*0.13*
Poland	3.2	*0.36*	13.8	*0.63*	27.5	*0.94*	29.4	*1.02*	19.3	*0.80*	6.1	*0.44*	0.7	*0.14*
Portugal	5.8	*0.76*	18.7	*1.05*	28.8	*0.92*	28.8	*1.22*	14.7	*0.88*	3.0	*0.35*	0.1	*0.05*
Slovak Republic	5.2	*0.60*	15.0	*0.87*	28.0	*0.96*	28.1	*0.99*	17.9	*1.02*	5.2	*0.49*	0.6	*0.14*
Spain	4.7	*0.44*	14.9	*0.69*	27.4	*0.77*	30.2	*0.68*	17.9	*0.75*	4.5	*0.38*	0.3	*0.10*
Sweden	3.8	*0.44*	12.6	*0.64*	25.2	*0.88*	29.5	*0.90*	21.1	*0.90*	6.8	*0.47*	1.1	*0.21*
Switzerland	4.5	*0.52*	11.6	*0.56*	21.8	*0.87*	28.2	*0.81*	23.5	*1.07*	9.1	*0.78*	1.4	*0.27*
Turkey	12.9	*0.83*	33.7	*1.31*	31.3	*1.42*	15.1	*1.06*	6.2	*1.15*	0.9	*0.32*	#	†
United Kingdom	4.8	*0.49*	11.9	*0.61*	21.8	*0.71*	25.9	*0.68*	21.8	*0.62*	10.9	*0.53*	2.9	*0.31*
United States	7.6	*0.94*	16.8	*0.88*	24.2	*0.94*	24.0	*0.79*	18.3	*0.97*	7.5	*0.62*	1.5	*0.25*
Non-OECD jurisdictions														
Argentina	28.3	*2.34*	27.9	*1.39*	25.6	*1.27*	13.6	*1.29*	4.1	*0.63*	0.4	*0.14*	#	†
Azerbaijan	19.4	*1.50*	53.1	*1.57*	22.4	*1.41*	4.7	*0.86*	0.4	*0.15*	#	†	#	†
Brazil	27.9	*0.99*	33.1	*0.96*	23.8	*0.93*	11.3	*0.88*	3.4	*0.42*	0.5	*0.21*	#	†
Bulgaria	18.3	*1.72*	24.3	*1.32*	25.2	*1.23*	18.8	*1.14*	10.3	*1.13*	2.6	*0.51*	0.4	*0.18*
Chile	13.1	*1.12*	26.7	*1.54*	29.9	*1.18*	20.1	*1.44*	8.4	*1.01*	1.8	*0.32*	0.1	*0.06*
Chinese Taipei	1.9	*0.29*	9.7	*0.82*	18.6	*0.86*	27.3	*0.80*	27.9	*1.03*	12.9	*0.77*	1.7	*0.24*
Colombia	26.2	*1.71*	34.0	*1.55*	27.2	*1.53*	10.6	*1.04*	1.9	*0.35*	0.2	*0.05*	#	†
Croatia	3.0	*0.43*	14.0	*0.71*	29.3	*0.91*	31.0	*0.99*	17.7	*0.86*	4.6	*0.44*	0.5	*0.12*
Estonia	1.0	*0.23*	6.7	*0.57*	21.0	*0.88*	33.7	*0.96*	26.2	*0.94*	10.1	*0.71*	1.4	*0.27*
Hong Kong-China	1.7	*0.36*	7.0	*0.68*	16.9	*0.81*	28.7	*0.95*	29.7	*0.95*	13.9	*0.80*	2.1	*0.30*
Indonesia	20.3	*1.71*	41.3	*2.23*	27.5	*1.46*	9.5	*1.99*	1.4	*0.53*	#	†	#	†
Israel	14.9	*1.18*	21.2	*1.01*	24.0	*0.95*	20.8	*0.96*	13.8	*0.80*	4.4	*0.49*	0.8	*0.18*
Jordan	16.2	*0.86*	28.2	*0.86*	30.8	*0.83*	18.7	*0.81*	5.6	*0.66*	0.6	*0.20*	#	†
Kyrgyz Republic	58.2	*1.56*	28.2	*1.13*	10.0	*0.81*	2.9	*0.39*	0.7	*0.18*	#	†	#	†
Latvia	3.6	*0.49*	13.8	*0.98*	29.0	*1.19*	32.9	*0.95*	16.6	*0.96*	3.8	*0.39*	0.3	*0.09*
Liechtenstein	2.6	*0.99*	10.3	*2.11*	21.0	*2.84*	28.7	*2.58*	25.2	*2.54*	10.0	*1.77*	2.2	*0.84*
Lithuania	4.3	*0.44*	16.0	*0.83*	27.4	*0.91*	29.8	*0.85*	17.5	*0.85*	4.5	*0.60*	0.4	*0.15*
Macao-China	1.4	*0.24*	8.9	*0.50*	26.0	*0.97*	35.7	*1.14*	22.8	*0.73*	5.0	*0.34*	0.3	*0.09*
Qatar	47.6	*0.62*	31.5	*0.63*	13.9	*0.49*	5.0	*0.35*	1.6	*0.14*	0.3	*0.09*	#	†
Republic of Montenegro	17.3	*0.79*	33.0	*1.20*	31.0	*0.91*	14.9	*0.65*	3.6	*0.37*	0.3	*0.11*	#	†
Republic of Serbia	11.9	*0.91*	26.6	*1.18*	32.3	*1.26*	21.8	*1.18*	6.6	*0.57*	0.8	*0.18*	#	†
Romania	16.0	*1.53*	30.9	*1.55*	31.8	*1.62*	16.6	*1.24*	4.2	*0.77*	0.5	*0.14*	#	†
Russian Federation	5.2	*0.65*	17.0	*1.08*	30.2	*0.93*	28.3	*1.32*	15.1	*1.09*	3.7	*0.46*	0.5	*0.13*
Slovenia	2.8	*0.34*	11.1	*0.72*	23.1	*0.68*	27.6	*1.08*	22.5	*1.13*	10.7	*0.57*	2.2	*0.29*
Thailand	12.6	*0.80*	33.5	*1.03*	33.2	*0.88*	16.3	*0.80*	4.0	*0.42*	0.4	*0.12*	#	†
Tunisia	27.7	*1.12*	35.1	*0.94*	25.0	*0.97*	10.2	*0.98*	1.9	*0.45*	0.1	*0.06*	#	†
Uruguay	16.7	*1.25*	25.4	*1.09*	29.8	*1.50*	19.7	*1.07*	6.9	*0.54*	1.3	*0.21*	0.1	*0.07*

† Not applicable.
# Rounds to zero.
NOTE: To reach a particular pro ciency level, a student must correctly answer a majority of items at that level. Students were classi ed into science literacy levels according to their scores. Exact cut point scores are as follows: below level 1 (a score less than or equal to 334.94); level 1 (a score greater than 334.94 and less than or equal to 409.54); level 2 (a score greater than 409.54 and less than or equal to 484.14); level 3 (a score greater than 484.14 and less than or equal to 558.73); level 4 (a score greater than 558.73 and less than or equal to 633.33); level 5 (a score greater than 633.33 and less than or equal to 707.93); and level 6 (a score greater than 707.93). The Organization for Economic Cooperation and Development (OECD) average is the average of the national averages of the OECD member jurisdictions. Because the Program for International Student Assessment (PISA) is principally an OECD study, the results for non-OECD jurisdictions are displayed separately from those of the OECD jurisdictions and are not included in the OECD average. Because of an error in printing the test booklets, the United States mean performance may be misestimated by approximately 1 score point. The impact is below one standard error. For details see appendix B. Standard error is noted by *s.e.* Detail may not sum to totals because of rounding.
SOURCE: Organization for Economic Cooperation and Development (OECD), Program for International Student Assessment (PISA), 2006.

Table C-6. Average scores of 15-year-old students on combined science literacy scale, by jurisdiction: 2000, 2003, and 2006

	2000		2003		2006	
Jurisdiction	**Average**	*s.e.*	**Average**	*s.e.*	**Average**	*s.e.*
OECD average	500	*0.7*	500	*0.6*	500	*0.5*
OECD jurisdictions						
Australia	528	*3.5*	525	*2.1*	527	*2.3*
Austria	519	*2.6*	491	*3.4*	511	*3.9*
Belgium	496	*4.3*	509	*2.4*	510	*2.5*
Canada	529	*1.6*	519	*2.0*	534	*2.0*
Czech Republic	511	*2.4*	523	*3.4*	513	*3.5*
Denmark	481	*2.8*	475	*3.0*	496	*3.1*
Finland	538	*2.5*	548	*1.9*	563	*2.0*
France	501	*3.2*	511	*3.0*	495	*3.4*
Germany	487	*2.4*	502	*3.6*	516	*3.8*
Greece	461	*4.9*	481	*3.8*	473	*3.2*
Hungary	496	*4.2*	503	*2.8*	504	*2.7*
Iceland	496	*2.2*	495	*1.5*	491	*1.6*
Ireland	513	*3.2*	505	*2.7*	508	*3.2*
Italy	478	*3.1*	487	*3.1*	475	*2.0*
Japan	550	*5.5*	548	*4.1*	531	*3.4*
Korea, Republic of	552	*2.7*	538	*3.5*	522	*3.4*
Luxembourg	443	*2.3*	483	*1.5*	486	*1.1*
Mexico	422	*3.2*	405	*3.5*	410	*2.7*
Netherlands[1]	—	—	524	*3.2*	525	*2.7*
New Zealand	528	*2.4*	521	*2.4*	530	*2.7*
Norway	500	*2.8*	484	*2.9*	487	*3.1*
Poland	483	*5.1*	498	*2.9*	498	*2.3*
Portugal	459	*4.0*	468	*3.5*	474	*3.0*
Slovak Republic	—	†	495	*3.7*	488	*2.6*
Spain	491	*3.0*	487	*2.6*	488	*2.6*
Sweden	512	*2.5*	506	*2.7*	503	*2.4*
Switzerland	496	*4.5*	513	*3.7*	512	*3.2*
Turkey	—	†	434	*5.9*	424	*3.8*
United Kingdom[2]	532	*2.7*	—	—	515	*2.3*
United States	500	*7.3*	491	*3.1*	489	*4.2*
Non-OECD jurisdictions						
Argentina	—	†	—	†	391	*6.1*
Azerbaijan	—	†	—	†	382	*2.8*
Brazil	—	†	390	*4.3*	390	*2.8*
Bulgaria	—	†	—	†	434	*6.1*
Chile	—	†	—	†	438	*4.3*
Chinese Taipei	—	†	—	†	532	*3.6*
Colombia	—	†	—	†	388	*3.4*
Croatia	—	†	—	†	493	*2.4*
Estonia	—	†	—	†	531	*2.5*
Hong Kong-China	—	†	540	*4.3*	542	*2.5*
Indonesia	—	†	395	*3.2*	393	*5.7*
Israel	—	†	—	†	454	*3.7*
Jordan	—	†	—	†	422	*2.8*
Kyrgyz Republic	—	†	—	†	322	*2.9*
Latvia	460	*5.6*	489	*3.9*	490	*3.0*
Liechtenstein	476	*7.1*	525	*4.3*	522	*4.1*
Lithuania	—	†	—	†	488	*2.8*
Macao-China	—	†	525	*3.0*	511	*1.1*
Qatar	—	†	—	†	349	*0.9*
Republic of Montenegro[3]	—	†	436	*3.5*	412	*1.1*
Republic of Serbia[3]	—	†	436	*3.5*	436	*3.0*

Table C-6. Continued

Jurisdiction	2000		2003		2006	
	Average	*s.e.*	Average	*s.e.*	Average	*s.e.*
OECD average	500	*0.7*	500	*0.6*	500	*0.5*
Romania	—	†	—	†	418	*4.2*
Russian Federation	460	*4.7*	489	*4.1*	479	*3.7*
Slovenia	—	†	—	†	519	*1.1*
Thailand	—	†	429	*2.7*	421	*2.1*
Tunisia	—	†	385	*2.6*	386	*3.0*
Uruguay	—	†	438	*2.9*	428	*2.7*

— Not available.

† Not applicable.

[1]Although the Netherlands participated in PISA in 2000, technical problems with its sample prevent its results from being discussed here.

[2] Because of low response rates, 2003 data for the United Kingdom are not discussed in this report.

[3] The Republics of Montenegro and Serbia were a united jurisdiction under the PISA 2003 assessment.

Note: The Organization for Economic Cooperation and Development (OECD) average is the average of the national averages of the OECD member jurisdictions with data available. Because the Program for International Student Assessment (PISA) is principally an OECD study, the results for non-OECD jurisdictions are displayed separately from those of the OECD jurisdictions and are not included in the OECD average. Because of an error in printing the test booklets, the United States mean performance in 2006 may be misestimated by approximately 1 score point. The impact is below one standard error. For details see appendix B. Standard error is noted by *s.e.*

Source: Organization for Economic Cooperation and Development (OECD), Program for International Student Assessment (PISA), 2000, 2003, and 2006.

Table C-7. Average scores of 15-year-old students on mathematics literacy scale, by jurisdiction: 2003 and 2006

Jurisdiction	2003		2006	
	Average	*s.e.*	Average	*s.e.*
OECD average	**500**	***0.6***	**498**	***0.5***
OECD jurisdictions				
Australia	524	*2.2*	520	*2.2*
Austria	506	*3.3*	505	*3.7*
Belgium	529	*2.3*	520	*3.0*
Canada	533	*1.8*	527	*2.0*
Czech Republic	517	*3.6*	510	*3.6*
Denmark	514	*2.7*	513	*2.6*
Finland	544	*1.9*	548	*2.3*
France	511	*2.5*	496	*3.2*
Germany	503	*3.3*	504	*3.9*
Greece	445	*3.9*	459	*3.0*
Hungary	490	*2.8*	491	*2.9*
Iceland	515	*1.4*	506	*1.8*
Ireland	503	*2.5*	501	*2.8*
Italy	466	*3.1*	462	*2.3*
Japan	534	*4.0*	523	*3.3*
Korea, Republic of	542	*3.2*	547	*3.8*
Luxembourg	493	*1.0*	490	*1.1*
Mexico	385	*3.6*	406	*2.9*
Netherlands	538	*3.1*	531	*2.6*
New Zealand	524	*2.3*	522	*2.4*
Norway	495	*2.4*	490	*2.6*
Poland	490	*2.5*	495	*2.4*

Table C-7. Continued

Jurisdiction	2003 Average	2003 s.e.	2006 Average	2006 s.e.
OECD average	**500**	***0.6***	**498**	***0.5***
Portugal	466	*3.4*	466	*3.1*
Slovak Republic	498	*3.4*	492	*2.8*
Spain	485	*2.4*	480	*2.3*
Sweden	509	*2.6*	502	*2.4*
Switzerland	527	*3.4*	530	*3.2*
Turkey	423	*6.7*	424	*4.9*
United Kingdom[1]	—	—	495	*2.1*
United States	483	*3.0*	474	*4.0*
Non-OECD jurisdictions				
Argentina	—	†	381	*6.2*
Azerbaijan	—	†	476	*2.3*
Brazil	356	*4.8*	370	*2.9*
Bulgaria	—	†	413	*6.1*
Chile	—	†	411	*4.6*
Chinese Taipei	—	†	549	*4.1*
Colombia	—	†	370	*3.8*
Croatia	—	†	467	*2.4*
Estonia	—	†	515	*2.7*
Hong Kong-China	550	*4.5*	547	*2.7*
Indonesia	360	*3.9*	391	*5.6*
Israel	—	†	442	*4.3*
Jordan	—	†	384	*3.3*
Kyrgyz Republic	—	†	311	*3.4*
Latvia	483	*3.7*	486	*3.0*
Liechtenstein	536	*4.1*	525	*4.2*
Lithuania	—	†	486	*2.9*
Macao-China	527	*2.9*	525	*1.3*
Qatar	—	†	318	*1.0*
Republic of Montenegro[2]	437	*3.8*	399	*1.4*
Republic of Serbia[2]	437	*3.8*	435	*3.5*
Romania	—	†	415	*4.2*
Russian Federation	468	*4.2*	476	*3.9*
Slovenia	—	†	504	*1.0*
Thailand	417	*3.0*	417	*2.3*
Tunisia	359	*2.5*	365	*4.0*
Uruguay	422	*3.3*	427	*2.6*

— Not available.

† Not applicable.

[1] Because of low response rates, 2003 data for the United Kingdom are not discussed in this report.

[2] The Republics of Montenegro and Serbia were a united jurisdiction under the PISA 2003 assessment.

Note: The Organization for Economic Cooperation and Development (OECD) average is the average of the national averages of the OECD member jurisdictions with data available. Because the Program for International Student Assessment (PISA) is principally an OECD study, the results for non-OECD jurisdictions are displayed separately from those of the OECD jurisdictions and are not included in the OECD average. Because of an error in printing the test booklets, the United States mean performance in 2006 may be misestimated by approximately 1 score point. The impact is below one standard error. For details see appendix B. Standard error is noted by *s.e.*

Source: Organization for Economic Cooperation and Development (OECD), Program for International Student Assessment (PISA), 2003 and 2006.

Table C-8. Scores of 15-year-old students on mathematics literacy scale at 10th and 90th percentiles, by jurisdiction: 2006

	Percentiles			
	10th		90th	
Jurisdiction	Score	*s.e.*	Score	*s.e.*
OECD average	**379**	***0.9***	**615**	***0.8***
OECD jurisdictions				
Australia	406	*2.7*	633	*3.3*
Austria	373	*6.3*	630	*3.8*
Belgium	381	*6.6*	650	*2.4*
Canada	416	*3.3*	635	*2.3*
Czech Republic	376	*4.7*	644	*4.8*
Denmark	404	*4.3*	621	*3.4*
Finland	444	*3.4*	652	*2.8*
France	369	*5.4*	617	*3.8*
Germany	375	*6.8*	632	*3.8*
Greece	341	*5.6*	575	*4.1*
Hungary	377	*3.9*	609	*5.0*
Iceland	391	*3.6*	618	*3.2*
Ireland	396	*4.4*	608	*3.2*
Italy	341	*3.3*	584	*4.2*
Japan	404	*5.5*	638	*3.6*
Korea, Republic of	426	*6.1*	664	*6.9*
Luxembourg	368	*3.5*	610	*2.7*
Mexico	299	*4.9*	514	*3.3*
Netherlands	412	*5.0*	645	*3.3*
New Zealand	401	*4.1*	643	*4.0*
Norway	373	*3.8*	609	*3.3*
Poland	384	*3.4*	610	*3.7*
Portugal	348	*5.2*	583	*2.8*
Slovak Republic	370	*5.1*	611	*4.4*
Spain	366	*2.8*	593	*2.9*
Sweden	387	*4.2*	617	*2.8*
Switzerland	401	*4.7*	652	*3.7*
Turkey	316	*4.0*	550	*12.4*
United Kingdom	381	*3.3*	612	*3.2*
United States	358	*5.8*	593	*4.8*
Non-OECD jurisdictions				
Argentina	249	*9.8*	508	*7.6*
Azerbaijan	419	*2.2*	536	*3.6*
Brazil	255	*4.5*	487	*5.8*
Bulgaria	287	*7.2*	543	*8.4*
Chile	302	*4.3*	527	*6.6*
Chinese Taipei	409	*6.2*	677	*3.4*
Colombia	258	*5.6*	482	*3.8*
Croatia	361	*3.3*	576	*3.6*
Estonia	411	*4.3*	618	*3.2*
Hong Kong-China	423	*6.4*	665	*3.5*
Indonesia	293	*3.9*	498	*9.4*
Israel	304	*6.9*	581	*5.0*
Jordan	279	*4.3*	489	*5.0*
Kyrgyz Republic	204	*5.0*	423	*5.9*
Latvia	378	*5.2*	590	*3.4*
Liechtenstein	402	*11.1*	643	*9.5*
Lithuania	369	*4.3*	602	*4.9*
Macao-China	416	*3.1*	632	*2.4*
Qatar	212	*2.2*	438	*2.7*
Republic of Montenegro	291	*3.0*	510	*2.4*

Table C-8. Continued

Jurisdiction	Percentiles 10th Score	s.e.	90th Score	s.e.
OECD average	**379**	***0.9***	**615**	***0.8***
Republic of Serbia	318	*5.0*	553	*3.9*
Romania	307	*7.4*	523	*7.1*
Russian Federation	363	*4.8*	592	*5.3*
Slovenia	390	*2.1*	623	*2.7*
Thailand	317	*3.5*	524	*3.7*
Tunisia	250	*3.9*	488	*7.8*
Uruguay	296	*4.4*	551	*5.5*

Note: The Organization for Economic Cooperation and Development (OECD) average is the average of the national averages of the OECD member jurisdictions. Because the Program for International Student Assessment (PISA) is principally an OECD study, the results for non-OECD jurisdictions are displayed separately from those of the OECD jurisdictions and are not included in the OECD average. Because of an error in printing the test booklets, the United States mean performance may be misestimated by approximately 1 score point. The impact is below one standard error. For details see appendix B. Standard error is noted by *s.e.*

Source: Organization for Economic Cooperation and Development (OECD), Program for International Student Assessment (PISA), 2006.

Table C-9. Average scores of 15-year-old students on combined science literacy scale, by sex and jurisdiction: 2006

Jurisdiction	Male Average	s.e.	Female Average	s.e.	Male-female difference Average	s.e.
OECD average	**501**	***0.7***	**499**	***0.6***	**2**	***0.7***
OECD jurisdictions						
Australia	527	*3.2*	527	*2.7*	#	†
Austria	515	*4.2*	507	*4.9*	8	*4.9*
Belgium	511	*3.3*	510	*3.2*	1	*4.1*
Canada	536	*2.5*	532	*2.1*	4	*2.2*
Czech Republic	515	*4.2*	510	*4.8*	5	*5.6*
Denmark	500	*3.6*	491	*3.4*	9	*3.2*
Finland	562	*2.6*	565	*2.4*	-3	*2.9*
France	497	*4.3*	494	*3.6*	3	*4.0*
Germany	519	*4.6*	512	*3.8*	7	*3.7*
Greece	468	*4.5*	479	*3.4*	-11	*4.7*
Hungary	507	*3.3*	501	*3.5*	6	*4.2*
Iceland	488	*2.6*	494	*2.1*	-6	*3.4*
Ireland	508	*4.3*	509	*3.3*	#	†
Italy	477	*2.8*	474	*2.5*	3	*3.5*
Japan	533	*4.9*	530	*5.1*	3	*7.4*
Korea, Republic of	521	*4.8*	523	*3.9*	-2	*5.5*
Luxembourg	491	*1.8*	482	*1.8*	9	*2.9*
Mexico	413	*3.2*	406	*2.6*	7	*2.2*
Netherlands	528	*3.2*	521	*3.1*	7	*3.0*
New Zealand	528	*3.9*	532	*3.6*	-4	*5.2*
Norway	484	*3.8*	489	*3.2*	-4	*3.4*
Poland	500	*2.7*	496	*2.6*	3	*2.5*
Portugal	477	*3.7*	472	*3.2*	5	*3.3*
Slovak Republic	491	*3.9*	485	*3.0*	6	*4.7*
Spain	491	*2.9*	486	*2.7*	4	*2.4*

Table C-9. Continued

Jurisdiction	Male		Female		Male-female difference	
	Average	*s.e.*	Average	*s.e.*	Average	*s.e.*
OECD average	**501**	***0.7***	**499**	***0.6***	**2**	***0.7***
Sweden	504	*2.7*	503	*2.9*	1	*3.0*
Switzerland	514	*3.3*	509	*3.6*	6	*2.7*
Turkey	418	*4.6*	430	*4.1*	-12	*4.1*
United Kingdom	520	*3.0*	510	*2.8*	10	*3.4*
United States	489	*5.1*	489	*4.0*	1	*3.5*
Non-OECD jurisdictions						
Argentina	384	*6.5*	397	*6.8*	-13	*5.6*
Azerbaijan	379	*3.1*	386	*2.7*	-8	*2.0*
Brazil	395	*3.2*	386	*2.9*	9	*2.3*
Bulgaria	426	*6.6*	443	*6.9*	-17	*5.8*
Chile	448	*5.4*	426	*4.4*	22	*4.8*
Chinese Taipei	536	*4.3*	529	*5.1*	7	*6.0*
Colombia	393	*4.1*	384	*4.1*	9	*4.6*
Croatia	492	*3.3*	494	*3.1*	-2	*4.1*
Estonia	530	*3.1*	533	*2.9*	-4	*3.1*
Hong Kong-China	546	*3.5*	539	*3.5*	7	*4.9*
Indonesia	399	*8.2*	387	*3.7*	12	*6.3*
Israel	456	*5.6*	452	*4.2*	3	*6.5*
Jordan	408	*4.5*	436	*3.3*	-29	*5.3*
Kyrgyz Republic	319	*3.6*	325	*3.0*	-6	*3.0*
Latvia	486	*3.5*	493	*3.2*	-7	*3.1*
Liechtenstein	516	*7.6*	527	*6.3*	-11	*11.1*
Lithuania	483	*3.1*	493	*3.1*	-9	*2.8*
Macao-China	513	*1.8*	509	*1.6*	4	*2.7*
Qatar	334	*1.2*	365	*1.3*	-32	*1.9*
Republic of Montenegro	411	*1.7*	413	*1.7*	-2	*2.6*
Republic of Serbia	433	*3.3*	438	*3.8*	-5	*3.8*
Romania	417	*4.1*	419	*4.8*	-2	*3.3*
Russian Federation	481	*4.1*	478	*3.7*	3	*2.7*
Slovenia	515	*2.0*	523	*1.9*	-8	*3.2*
Thailand	411	*3.4*	428	*2.5*	-17	*3.9*
Tunisia	383	*3.2*	388	*3.5*	-5	*3.4*
Uruguay	427	*4.0*	430	*2.7*	-3	*4.0*

† Not applicable. # Rounds to zero. NOTE: The Organization for Economic Cooperation and Development (OECD) average is the average of the national averages of the OECD member jurisdictions. Because the Program for International Student Assessment (PISA) is principally an OECD study, the results for non-OECD jurisdictions are displayed separately from those of the OECD jurisdictions and are not included in the OECD average. Differences were computed using unrounded numbers. Because of an error in printing the test booklets, the United States mean performance may be misestimated by approximately 1 score point. The impact is below one standard error. For details see appendix B. Standard error is noted by *s.e.*

Source: Organization for Economic Cooperation and Development (OECD), Program for International Student Assessment (PISA), 2006.

Table C-10. Percentage distribution of 15-year-old students at each proficiency level on combined science literacy scale, by sex and jurisdiction: 2006

Jurisdiction	Below level 1				Level 1				Level 2				Level 3			
	Male	Female	Male	Female	Male	Female	Male	Female	Male	Female	Male	Female	Male	Female	Male	Female
	Percent	*s.e.*	Percent	*s.e.*	Percent	*s.e.*	Percent	*s.e.*	Percent	*s.e.*	Percent	*s.e.*	Percent	*s.e.*	Percent	*s.e.*
OECD average	**5.6**	***0.15***	**4.7**	***0.13***	**14.1**	***0.19***	**14.0**	***0.19***	**23.4**	***0.23***	**24.7**	***0.23***	**26.4**	***0.22***	**28.5**	***0.23***
OECD jurisdictions																
Australia	3.6	*0.35*	2.5	*0.33*	10.3	*0.58*	9.3	*0.60*	19.7	*0.85*	20.8	*0.80*	26.6	*0.91*	28.9	*0.62*
Austria	3.6	*0.76*	5.0	*1.30*	11.6	*1.26*	12.5	*1.18*	22.7	*1.60*	20.8	*1.54*	27.5	*1.35*	29.0	*1.52*
Belgium	5.0	*1.03*	4.6	*0.66*	12.9	*1.00*	11.4	*0.81*	20.8	*1.05*	20.8	*1.14*	25.6	*0.91*	29.9	*1.40*
Canada	2.4	*0.37*	1.9	*0.33*	8.1	*0.68*	7.5	*0.69*	18.1	*0.70*	20.0	*0.85*	27.5	*0.74*	30.2	*0.91*
Czech Republic	2.6	*0.47*	4.7	*0.95*	11.7	*1.02*	12.5	*1.18*	24.5	*1.60*	22.0	*1.45*	28.0	*1.40*	27.5	*1.40*
Denmark	4.2	*0.66*	4.5	*0.83*	13.6	*1.00*	14.5	*1.03*	24.8	*1.24*	27.1	*1.27*	28.6	*1.16*	30.0	*1.43*
Finland	0.6	*0.21*	0.4	*0.17*	4.3	*0.61*	2.8	*0.49*	14.6	*0.83*	12.6	*0.91*	28.0	*1.26*	30.3	*1.27*
France	7.5	*0.96*	5.8	*0.74*	14.5	*1.18*	14.6	*1.23*	22.2	*1.39*	23.4	*1.39*	25.3	*1.50*	28.9	*1.34*
Germany	4.4	*0.84*	3.7	*0.67*	10.5	*1.09*	12.1	*1.19*	21.6	*1.23*	21.1	*1.26*	25.9	*1.21*	29.9	*1.47*
Greece	9.3	*1.28*	5.1	*0.81*	18.9	*1.29*	14.9	*0.95*	27.2	*1.24*	30.7	*1.83*	26.4	*1.37*	32.5	*1.53*
Hungary	2.8	*0.50*	2.6	*0.49*	12.8	*1.09*	11.9	*1.16*	25.2	*1.45*	26.9	*1.61*	28.7	*1.28*	33.6	*1.74*
Iceland	6.9	*0.69*	4.7	*0.66*	15.5	*1.02*	14.0	*1.11*	25.8	*1.37*	25.9	*1.23*	26.0	*1.61*	30.5	*1.46*
Ireland	4.1	*0.68*	3.0	*0.51*	12.5	*1.28*	11.5	*0.91*	23.2	*1.24*	24.8	*1.68*	28.8	*1.22*	30.6	*1.58*
Italy	8.0	*0.71*	6.5	*0.52*	17.5	*0.88*	18.5	*0.83*	25.9	*0.96*	29.3	*0.98*	27.4	*0.86*	27.4	*0.91*
Japan	3.6	*0.59*	2.8	*0.72*	9.2	*1.03*	8.5	*1.04*	18.1	*1.07*	18.8	*1.23*	25.8	*1.07*	29.2	*1.25*
Korea, Republic of	3.2	*0.70*	1.8	*0.45*	9.2	*0.99*	8.3	*1.05*	20.8	*1.57*	21.5	*1.13*	30.2	*1.42*	33.3	*1.43*
Luxembourg	7.0	*0.57*	6.1	*0.58*	15.1	*1.02*	16.1	*1.01*	23.8	*1.25*	27.0	*1.07*	27.2	*1.31*	29.9	*1.10*
Mexico	17.4	*1.55*	18.9	*1.28*	32.1	*1.27*	33.4	*1.10*	30.5	*1.39*	31.0	*1.08*	15.8	*0.85*	13.9	*0.76*
Netherlands	2.4	*0.47*	2.2	*0.49*	9.9	*0.99*	11.5	*1.16*	20.7	*1.38*	21.6	*1.21*	27.3	*1.23*	26.6	*1.25*
New Zealand	5.0	*0.69*	3.1	*0.45*	10.3	*0.84*	9.1	*0.77*	19.4	*1.16*	20.0	*1.18*	24.1	*1.17*	26.0	*0.96*
Norway	7.3	*1.19*	4.3	*0.68*	15.1	*0.91*	15.3	*1.12*	26.5	*1.13*	28.1	*1.08*	27.7	*1.12*	29.4	*1.57*
Poland	3.7	*0.54*	2.7	*0.43*	13.6	*0.77*	13.9	*0.83*	26.9	*1.51*	28.1	*1.04*	28.6	*1.36*	30.3	*1.21*
Portugal	5.9	*0.93*	5.6	*0.88*	18.3	*1.53*	19.0	*1.14*	28.3	*1.19*	29.3	*1.17*	27.9	*1.46*	29.8	*1.49*
Slovak Republic	5.5	*0.94*	4.8	*0.70*	14.6	*1.07*	15.5	*1.33*	27.0	*1.45*	29.2	*1.23*	27.4	*1.45*	28.8	*1.32*
Spain	5.2	*0.54*	4.3	*0.55*	14.4	*0.89*	15.4	*0.87*	26.4	*1.03*	28.3	*1.19*	29.7	*0.99*	30.7	*0.89*
Sweden	4.1	*0.62*	3.4	*0.48*	13.1	*0.90*	12.0	*0.86*	24.0	*1.14*	26.4	*1.56*	28.6	*1.40*	30.4	*1.44*
Switzerland	4.6	*0.63*	4.4	*0.52*	10.9	*0.64*	12.2	*0.80*	20.8	*1.06*	22.8	*1.06*	28.5	*1.06*	27.8	*1.04*
Turkey	15.2	*1.21*	10.1	*1.12*	35.0	*1.60*	32.2	*1.79*	29.0	*1.60*	34.1	*1.86*	13.8	*1.16*	16.6	*1.39*
United Kingdom	5.3	*0.74*	4.3	*0.48*	11.4	*0.90*	12.4	*0.88*	20.5	*0.80*	23.0	*1.03*	24.1	*0.86*	27.7	*1.02*
United States	8.3	*1.23*	6.8	*0.85*	17.4	*1.28*	16.2	*1.06*	22.3	*1.18*	26.2	*1.16*	23.4	*1.10*	24.6	*1.02*

Table C-10. Continued

Jurisdiction	Below level 1 Male Percent	Female s.e.	Male Percent	Female s.e.	Level 1 Male Percent	Female s.e.	Male Percent	Female s.e.	Level 2 Male Percent	Female s.e.	Male Percent	Female s.e.	Level 3 Male Percent	Female s.e.	Male Percent	Female s.e.
OECD average	**5.6**	***0.15***	**4.7**	***0.13***	**14.1**	***0.19***	**14.0**	***0.19***	**23.4**	***0.23***	**24.7**	***0.23***	**26.4**	***0.22***	**28.5**	***0.23***
Non-OECD jurisdictions																
Argentina	30.7	*2.65*	26.2	*2.45*	28.1	*1.74*	27.8	*1.66*	25.1	*1.61*	26.0	*1.52*	12.2	*1.30*	14.9	*1.70*
Azerbaijan	22.4	*1.82*	16.1	*1.67*	52.2	*1.83*	54.1	*1.81*	20.2	*1.61*	24.8	*1.77*	4.7	*0.96*	4.7	*0.95*
Brazil	26.8	*1.17*	28.9	*1.21*	31.6	*1.24*	34.4	*1.11*	24.9	*1.16*	22.8	*1.20*	11.9	*1.17*	10.7	*0.88*
Bulgaria	21.2	*2.14*	15.2	*1.80*	25.5	*1.57*	23.1	*1.86*	23.4	*1.59*	27.0	*1.76*	17.3	*1.33*	20.5	*1.61*
Chile	10.8	*1.17*	15.7	*1.42*	25.0	*1.95*	28.6	*1.54*	29.7	*1.53*	30.1	*1.46*	22.2	*1.54*	17.6	*1.62*
Chinese Taipei	2.0	*0.38*	1.9	*0.41*	9.7	*0.97*	9.7	*1.06*	17.4	*0.92*	19.9	*1.38*	26.4	*1.21*	28.3	*1.13*
Colombia	25.2	*1.91*	27.0	*2.01*	32.2	*1.76*	35.5	*2.12*	27.6	*1.98*	26.9	*1.96*	12.3	*1.52*	9.1	*1.21*
Croatia	3.4	*0.68*	2.6	*0.50*	14.8	*1.02*	13.1	*1.18*	28.8	*1.18*	29.7	*1.47*	29.7	*1.17*	32.3	*1.44*
Estonia	1.2	*0.37*	0.7	*0.21*	7.4	*0.76*	6.0	*0.68*	21.0	*1.11*	21.0	*1.15*	33.2	*1.18*	34.2	*1.45*
Hong Kong-China	1.9	*0.54*	1.5	*0.33*	7.3	*0.85*	6.7	*0.83*	15.9	*1.09*	17.9	*1.13*	26.8	*1.13*	30.5	*1.56*
Indonesia	18.7	*2.18*	22.0	*1.61*	39.9	*3.15*	42.7	*1.95*	28.0	*2.01*	27.0	*1.66*	11.5	*3.04*	7.3	*1.21*
Israel	16.0	*1.62*	13.8	*1.41*	21.3	*1.33*	21.1	*1.14*	21.7	*1.07*	26.3	*1.42*	19.6	*1.09*	22.0	*1.37*
Jordan	21.6	*1.40*	10.8	*1.03*	29.2	*1.36*	27.1	*1.17*	27.8	*1.22*	33.7	*1.04*	16.2	*1.21*	21.2	*1.28*
Kyrgyz Republic	60.0	*1.85*	56.6	*1.69*	26.2	*1.31*	29.9	*1.30*	9.7	*0.97*	10.4	*0.93*	3.2	*0.60*	2.6	*0.43*
Latvia	4.0	*0.64*	3.2	*0.61*	15.1	*1.14*	12.7	*1.15*	29.3	*1.62*	28.7	*1.51*	31.9	*1.52*	33.9	*1.32*
Liechtenstein	3.0	*1.73*	2.3	*1.22*	10.2	*3.89*	10.3	*2.63*	22.8	*4.58*	19.4	*3.56*	31.0	*4.43*	26.7	*3.35*
Lithuania	4.9	*0.55*	3.8	*0.57*	17.2	*0.99*	14.8	*1.26*	27.9	*1.24*	26.8	*1.38*	28.5	*1.20*	31.1	*1.25*
Macao-China	1.8	*0.32*	1.0	*0.27*	9.5	*0.70*	8.2	*0.67*	24.2	*0.99*	27.8	*1.41*	34.4	*1.52*	36.9	*1.54*
Qatar	57.7	*0.96*	37.3	*0.92*	26.2	*0.99*	36.9	*0.96*	9.5	*0.72*	18.3	*0.88*	4.3	*0.38*	5.9	*0.57*
Republic of Montenegro	17.7	*0.95*	16.8	*1.15*	33.1	*1.55*	32.8	*1.77*	30.4	*1.43*	31.6	*1.11*	15.0	*0.91*	14.8	*0.91*
Republic of Serbia	12.9	*1.09*	10.9	*1.19*	27.9	*1.44*	25.3	*1.48*	31.0	*1.56*	33.5	*1.52*	20.6	*1.48*	23.1	*1.43*
Romania	17.6	*1.57*	14.3	*1.93*	30.7	*1.73*	31.2	*1.96*	29.4	*1.56*	34.2	*2.46*	16.9	*1.63*	16.2	*1.54*
Russian Federation	5.6	*0.77*	4.9	*0.72*	17.0	*1.15*	16.9	*1.35*	29.3	*1.35*	31.1	*1.09*	27.5	*1.97*	29.1	*1.28*
Slovenia	3.2	*0.40*	2.4	*0.51*	12.1	*0.97*	10.1	*0.75*	24.0	*1.14*	22.3	*0.86*	26.6	*1.56*	28.6	*1.13*
Thailand	17.1	*1.56*	9.3	*0.92*	34.7	*1.41*	32.6	*1.19*	29.1	*1.32*	36.3	*1.18*	14.9	*1.05*	17.3	*1.08*
Tunisia	29.3	*1.43*	26.2	*1.36*	34.2	*1.27*	35.9	*1.40*	24.5	*1.20*	25.4	*1.30*	9.8	*1.19*	10.6	*1.27*
Uruguay	18.2	*1.84*	15.3	*1.26*	25.8	*1.85*	25.0	*1.19*	27.8	*1.92*	31.7	*1.93*	18.9	*1.45*	20.5	*1.46*

Table C-10. Continued

Jurisdiction	Level 4				Level 5				Level 6			
	Male Percent	Female s.e.	Male Percent	Female s.e.	Male Percent	Female s.e.	Male Percent	Female s.e.	Male Percent	Female s.e.	Male Percent	Female s.e.
OECD average	**20.5**	***0.21***	**20.2**	***0.21***	**8.5**	***0.15***	**6.9**	***0.13***	**1.5**	***0.06***	**1.0**	***0.05***
OECD jurisdictions												
Australia	24.2	*0.72*	25.0	*0.71*	12.3	*0.73*	11.2	*0.68*	3.3	*0.43*	2.4	*0.29*
Austria	23.3	*1.38*	24.1	*1.62*	9.7	*0.97*	7.9	*0.89*	1.6	*0.30*	0.8	*0.21*
Belgium	24.5	*0.92*	24.4	*1.21*	9.9	*0.67*	8.3	*0.70*	1.3	*0.24*	0.6	*0.22*
Canada	28.1	*0.94*	27.2	*0.85*	12.9	*0.61*	11.2	*0.77*	2.8	*0.29*	2.0	*0.32*
Czech Republic	21.4	*1.42*	22.2	*1.27*	9.9	*1.02*	9.6	*1.10*	2.0	*0.41*	1.6	*0.35*
Denmark	21.0	*1.24*	18.1	*1.12*	7.0	*0.92*	5.2	*0.68*	0.8	*0.26*	0.6	*0.28*
Finland	30.8	*1.07*	33.7	*1.16*	17.0	*0.96*	16.9	*0.97*	4.6	*0.50*	3.3	*0.48*
France	20.9	*1.29*	20.8	*1.26*	8.5	*0.80*	6.0	*0.83*	1.1	*0.33*	0.5	*0.16*
Germany	23.8	*1.35*	23.3	*1.07*	11.5	*1.03*	8.4	*0.73*	2.2	*0.37*	1.4	*0.38*
Greece	14.2	*1.06*	14.1	*1.09*	3.7	*0.50*	2.7	*0.46*	0.3	*0.14*	0.1	*0.08*
Hungary	22.0	*1.14*	19.8	*1.28*	7.6	*0.94*	4.8	*0.74*	0.8	*0.21*	0.4	*0.20*
Iceland	19.2	*1.11*	18.8	*1.00*	5.8	*0.71*	5.4	*0.75*	0.8	*0.22*	0.7	*0.29*
Ireland	21.1	*1.10*	21.6	*1.23*	8.9	*0.92*	7.6	*0.75*	1.4	*0.31*	0.9	*0.29*
Italy	15.8	*0.73*	14.4	*0.73*	4.9	*0.44*	3.6	*0.41*	0.6	*0.13*	0.3	*0.09*
Japan	26.5	*1.51*	27.5	*1.58*	13.7	*0.93*	11.2	*0.91*	3.3	*0.48*	2.0	*0.35*
Korea, Republic of	25.5	*1.32*	25.5	*1.31*	9.9	*1.13*	8.6	*0.89*	1.3	*0.37*	0.9	*0.33*
Luxembourg	19.6	*1.07*	16.5	*0.91*	6.6	*0.63*	4.1	*0.53*	0.8	*0.23*	0.3	*0.16*
Mexico	3.8	*0.41*	2.6	*0.36*	0.3	*0.12*	0.2	*0.10*	#	*t*	#	*t*
Netherlands	24.9	*1.30*	26.8	*1.27*	13.0	*1.13*	9.9	*0.75*	2.0	*0.44*	1.3	*0.31*
New Zealand	22.8	*1.10*	24.9	*1.07*	14.0	*0.98*	13.3	*1.05*	4.4	*0.67*	3.6	*0.50*
Norway	16.7	*1.22*	17.5	*1.20*	6.0	*0.68*	4.9	*0.69*	0.7	*0.20*	0.5	*0.18*
Poland	19.1	*1.09*	19.5	*1.06*	7.2	*0.65*	5.0	*0.62*	0.9	*0.25*	0.5	*0.18*
Portugal	15.5	*1.00*	14.0	*1.20*	3.9	*0.60*	2.2	*0.31*	0.1	*0.09*	#	*t*
Slovak Republic	18.8	*1.35*	17.0	*1.32*	6.0	*0.79*	4.4	*0.55*	0.8	*0.29*	0.4	*0.18*
Spain	18.7	*0.95*	17.1	*0.87*	5.1	*0.49*	4.0	*0.44*	0.5	*0.16*	0.2	*0.08*
Sweden	21.5	*1.13*	20.6	*1.27*	7.3	*0.69*	6.2	*0.75*	1.2	*0.34*	1.0	*0.26*
Switzerland	24.0	*1.24*	23.0	*1.27*	9.7	*0.87*	8.4	*0.95*	1.4	*0.33*	1.4	*0.35*
Turkey	6.2	*1.28*	6.1	*1.23*	0.9	*0.37*	0.9	*0.40*	#	*t*	#	*t*
United Kingdom	22.5	*0.81*	21.1	*0.97*	12.3	*0.78*	9.4	*0.71*	3.7	*0.48*	2.1	*0.39*
United States	18.6	*1.33*	18.0	*1.01*	8.4	*0.84*	6.7	*0.78*	1.6	*0.30*	1.5	*0.35*

Table C-10. Continued

	Level 4				Level 5				Level 6			
	Male	Female	Male	Female	Male	Female	Male	Female	Male	Female	Male	Female
Jurisdiction	Percent	*s.e.*	Percent	*s.e.*	Percent	*s.e.*	Percent	*s.e.*	Percent	*s.e.*	Percent	*s.e.*
OECD average	**20.5**	***0.21***	**20.2**	***0.21***	**8.5**	***0.15***	**6.9**	***0.13***	**1.5**	***0.06***	**1.0**	***0.05***
Non-OECD jurisdictions												
Argentina	3.4	*0.58*	4.6	*0.92*	0.4	*0.16*	0.5	*0.22*	#	†	#	†
Azerbaijan	0.4	*0.15*	0.4	*0.19*	#	†	#	†	#	†	#	†
Brazil	4.0	*0.55*	2.8	*0.50*	0.7	*0.31*	0.4	*0.18*	0.1	*0.06*	#	†
Bulgaria	9.2	*1.18*	11.4	*1.49*	2.8	*0.62*	2.4	*0.52*	0.5	*0.23*	0.4	*0.18*
Chile	9.9	*1.29*	6.6	*1.02*	2.3	*0.54*	1.2	*0.42*	0.1	*0.08*	0.1	*0.11*
Chinese Taipei	28.8	*1.18*	26.9	*1.51*	13.8	*1.08*	12.0	*1.13*	2.0	*0.42*	1.4	*0.30*
Colombia	2.5	*0.53*	1.4	*0.42*	0.2	*0.11*	0.1	*0.09*	#	†	#	†
Croatia	17.9	*1.01*	17.5	*1.24*	4.7	*0.52*	4.4	*0.63*	0.7	*0.16*	0.4	*0.15*
Estonia	25.4	*1.40*	27.0	*1.25*	10.2	*0.90*	10.0	*0.98*	1.6	*0.32*	1.2	*0.33*
Hong Kong-China	30.4	*1.27*	29.1	*1.25*	14.7	*1.06*	13.0	*1.17*	2.8	*0.50*	1.3	*0.27*
Indonesia	1.8	*0.84*	1.0	*0.40*	#	†	#	†	#	†	#	†
Israel	14.7	*1.25*	12.9	*0.96*	5.4	*0.75*	3.5	*0.47*	1.3	*0.28*	0.3	*0.16*
Jordan	4.6	*0.88*	6.5	*0.75*	0.6	*0.26*	0.7	*0.21*	#	†	#	†
Kyrgyz Republic	1.0	*0.30*	0.5	*0.16*	#	†	#	†	#	†	#	†
Latvia	15.4	*1.24*	17.7	*1.18*	4.0	*0.60*	3.7	*0.46*	0.3	*0.13*	0.2	*0.11*
Liechtenstein	20.8	*4.08*	29.0	*3.72*	10.6	*2.79*	9.5	*2.34*	1.5	*1.17*	2.8	*1.35*
Lithuania	16.9	*1.09*	18.1	*1.18*	4.1	*0.62*	5.0	*0.76*	0.4	*0.21*	0.4	*0.20*
Macao-China	23.5	*1.60*	22.0	*1.11*	6.2	*0.58*	3.8	*0.56*	0.3	*0.18*	0.2	*0.12*
Qatar	1.9	*0.24*	1.4	*0.22*	0.4	*0.14*	0.2	*0.10*	#	†	#	†
Republic of Montenegro	3.5	*0.50*	3.8	*0.54*	0.3	*0.16*	0.2	*0.17*	#	†	#	†
Republic of Serbia	6.5	*0.71*	6.6	*0.69*	1.0	*0.27*	0.6	*0.21*	#	†	#	†
Romania	4.6	*0.83*	3.9	*1.02*	0.7	*0.24*	0.2	*0.13*	#	†	#	†
Russian Federation	15.6	*1.45*	14.6	*1.12*	4.4	*0.69*	3.0	*0.42*	0.7	*0.21*	0.3	*0.15*
Slovenia	21.5	*1.54*	23.5	*1.44*	10.2	*0.95*	11.2	*1.00*	2.4	*0.52*	1.9	*0.38*
Thailand	3.8	*0.62*	4.1	*0.55*	0.5	*0.20*	0.4	*0.14*	#	†	#	†
Tunisia	2.0	*0.47*	1.8	*0.58*	0.1	*0.09*	0.1	*0.12*	#	†	#	†
Uruguay	7.3	*0.67*	6.5	*0.67*	1.7	*0.35*	0.9	*0.30*	0.2	*0.12*	0.1	*0.06*

† Not applicable.

Rounds to zero.

Note: To reach a particular proficiency level, a student must correctly answer a majority of items at that level. Students were classified into science literacy levels according totheir scores. Exact cut point scores are as follows: below level 1 (a score less than or equal to 334.94); level 1 (a score greater than 334.94 and less than or equal to 409.54); level 2 (a score greater than 409.54 and less than or equal to 484.14); level 3 (a score greater than 484.14 and less than or equal to 558.73); level 4 (a score greater than 558.73 and less than or equal to 633.33); level 5 (a score greater than 633.33 and less than or equal to

707.93); and level 6 (a score greater than 707.93). The Organization for Economic Cooperation and Development (OECD) average is the average of the national averages of the OECD member jurisdictions. Because the Program for International Student Assessment (PISA) is principally an OECD study, the results for non-OECD jurisdictions are displayed separately from those of the OECD jurisdictions and are not included in the OECD average. Because of an error in printing the test booklets, the United States mean performance may be misestimated by approximately 1 score point. The impact is below one standard error. For details see appendix B. Standard error is noted by *s.e.* Detail may not sum to totals because of rounding.

Source: Organization for Economic Cooperation and Development (OECD), Program for International Student Assessment (PISA), 2006.

Table C-11. Average scores of 15-year-old students on science literacy subscales, by sex and jurisdiction: 2006

Jurisdiction	Identiffying Scien tific Issues Male Aver.	s.e.	Female Aver	s.e.	Male-Female Difference Aver	s.e.	Explaining Phenomenas Scientifically Male Aver.	s.e.	Female Aver	s.e.	Male-Female Aver	s.e.
OECD average	**490**	***0.7***	**508**	***0.6***	**-17**	***0.7***	**508**	***0.7***	**493**	***0.6***	**15**	***0.7***
OECD jurisdictions												
Australia	525	*3.2*	546	*2.6*	-21	*3.6*	527	*3.1*	513	*2.7*	13	*3.6*
Austria	495	*4.2*	516	*4.7*	-22	*4.6*	526	*4.4*	507	*4.7*	19	*4.8*
Belgium	508	*3.8*	523	*3.1*	-14	*4.3*	510	*3.4*	494	*3.1*	16	*4.1*
Canada	525	*2.7*	539	*2.4*	-14	*2.4*	539	*2.6*	522	*2.3*	17	*2.5*
Czech Republic	492	*4.8*	511	*5.3*	-19	*5.7*	537	*4.3*	516	*4.6*	21	*5.7*
Denmark	488	*3.5*	499	*3.2*	-11	*3.2*	512	*3.8*	491	*3.7*	21	*3.4*
Finland	542	*2.7*	568	*2.6*	-26	*2.8*	571	*2.5*	562	*2.5*	9	*3.0*
France	491	*4.6*	507	*3.7*	-16	*4.7*	489	*4.2*	474	*3.4*	15	*4.1*
Germany	502	*4.5*	518	*3.9*	-16	*3.4*	529	*4.5*	508	*3.7*	21	*3.7*
Greece	453	*4.1*	485	*3.1*	-31	*4.3*	478	*4.3*	475	*3.0*	3	*4.2*
Hungary	477	*3.4*	489	*3.3*	-13	*4.1*	529	*3.2*	507	*3.6*	22	*4.4*
Iceland	479	*2.9*	509	*2.4*	-30	*4.1*	491	*2.6*	485	*2.1*	6	*3.7*
Ireland	508	*4.4*	524	*3.5*	-16	*4.6*	510	*4.4*	501	*3.5*	9	*4.6*
Italy	466	*2.9*	483	*2.5*	-17	*3.4*	487	*2.8*	472	*2.5*	15	*3.4*
Japan	513	*5.1*	531	*6.6*	-18	*8.5*	535	*4.6*	519	*4.4*	16	*6.6*
Korea, Republic of	508	*4.9*	530	*4.2*	-22	*5.7*	517	*4.8*	506	*4.0*	11	*5.7*
Luxembourg	477	*1.7*	489	*1.8*	-11	*2.8*	495	*1.8*	471	*2.0*	25	*3.0*
Mexico	418	*2.9*	425	*2.8*	-7	*2.2*	415	*3.3*	398	*2.6*	18	*2.3*
Netherlands	527	*3.8*	539	*3.5*	-12	*3.2*	531	*3.1*	512	*3.1*	18	*3.0*
New Zealand	525	*3.7*	547	*3.7*	-22	*4.9*	528	*4.0*	517	*3.6*	11	*5.2*
Norway	478	*3.9*	501	*3.3*	-24	*3.7*	498	*3.9*	492	*3.2*	6	*3.9*
Poland	476	*2.8*	490	*2.7*	-13	*2.5*	514	*2.9*	498	*2.8*	17	*2.7*
Portugal	480	*3.6*	493	*3.4*	-13	*3.1*	477	*3.6*	462	*3.0*	16	*3.2*
Slovak Republic	465	*4.5*	485	*3.6*	-20	*5.1*	512	*4.0*	490	*3.0*	22	*4.7*
Spain	482	*2.7*	496	*2.6*	-15	*2.1*	499	*2.8*	481	*2.7*	18	*2.6*
Sweden	491	*2.9*	507	*3.1*	-16	*3.0*	516	*3.0*	504	*3.5*	12	*3.1*
Switzerland	510	*3.1*	520	*3.3*	-10	*2.4*	517	*3.4*	498	*3.9*	18	*2.8*
Turkey	414	*4.1*	443	*3.6*	-29	*3.8*	423	*4.7*	423	*4.5*	1	*4.1*
United Kingdom	510	*2.9*	517	*2.8*	-7	*3.2*	527	*3.0*	506	*2.7*	21	*3.5*
United States	484	*4.6*	500	*3.8*	-16	*3.6*	492	*5.3*	480	*4.0*	13	*3.6*
Non-OECD jurisdictions												
Argentina	381	*5.8*	408	*6.4*	-27	*5.2*	387	*6.4*	386	*7.0*	0	*5.8*
Azerbaijan	349	*3.3*	357	*3.3*	-8	*2.3*	408	*3.3*	417	*3.0*	-9	*1.9*
Brazil	394	*3.2*	402	*3.0*	-7	*2.5*	400	*3.0*	382	*2.9*	19	*2.4*
Bulgaria	411	*6.6*	445	*7.1*	-34	*5.6*	442	*6.5*	447	*6.5*	-5	*5.8*
Chile	445	*5.0*	443	*4.1*	3	*4.5*	448	*5.1*	414	*4.1*	34	*4.6*
Chinese Taipei	506	*4.4*	512	*5.0*	-6	*5.8*	554	*4.3*	535	*5.3*	19	*6.1*
Colombia	401	*4.4*	404	*4.0*	-3	*4.8*	388	*4.3*	371	*4.3*	18	*4.8*
Croatia	480	*3.5*	507	*3.1*	-27	*4.1*	498	*3.2*	487	*3.3*	11	*4.1*
Estonia	504	*3.1*	528	*2.6*	-25	*2.8*	544	*3.2*	537	*3.0*	6	*3.3*
Hong Kong-China	520	*4.1*	535	*4.5*	-15	*5.9*	560	*3.5*	539	*3.3*	21	*4.6*
Indonesia	397	*8.0*	389	*3.6*	8	*6.0*	403	*7.0*	386	*3.8*	17	*5.7*
Israel	451	*5.9*	463	*4.0*	-12	*6.6*	451	*5.4*	436	*4.0*	16	*6.4*
Jordan	393	*4.6*	425	*2.8*	-32	*5.1*	427	*4.6*	448	*4.1*	-21	*6.0*
Kyrgyz Republic	311	*3.6*	330	*3.3*	-20	*2.9*	335	*3.9*	333	*2.9*	2	*3.0*
Latvia	473	*3.7*	504	*3.5*	-31	*3.1*	491	*3.6*	481	*3.2*	10	*3.3*
Liechtenstein	508	*7.0*	534	*5.7*	-26	*10.3*	519	*7.5*	513	*6.4*	6	*11.1*
Lithuania	463	*2.9*	489	*3.0*	-26	*2.7*	499	*3.3*	490	*3.4*	9	*3.1*
Macao-China	483	*1.9*	498	*1.6*	-15	*2.6*	527	*2.0*	513	*1.6*	14	*2.7*

Table C-11. Continued

Jurisdiction	Identiffying Scien tific Issues						Explaining Phenomenas Scientifically					
	Male		Female		Male-Female Difference		Male		Female		Male-Female	
	Aver.	s.e.	Aver	s.e.	Aver	s.e.	Aver.	s.e.	Aver	s.e.	Aver	s.e.
OECD average	**490**	***0.7***	**508**	***0.6***	**-17**	***0.7***	**508**	***0.7***	**493**	***0.6***	**15**	***0.7***
Qatar	334	*1.2*	371	*1.3*	-37	*2.1*	342	*1.4*	371	*1.6*	-29	*2.3*
Rep. of Montenegro	393	*2.0*	409	*1.8*	-16	*2.9*	421	*1.8*	412	*1.7*	9	*2.7*
Republic of Serbia	420	*3.3*	441	*3.6*	-21	*3.7*	444	*3.7*	438	*3.8*	6	*4.1*
Romania	401	*3.6*	418	*4.4*	-17	*3.5*	431	*4.3*	421	*4.5*	10	*3.6*
Russian Federation	453	*4.6*	472	*4.1*	-20	*2.6*	493	*4.0*	474	*3.4*	19	*2.6*
Slovenia	504	*2.0*	530	*2.0*	-27	*2.8*	528	*2.3*	518	*2.2*	10	*3.3*
Thailand	394	*3.7*	427	*2.8*	-33	*4.1*	418	*3.4*	421	*2.2*	-3	*3.6*
Tunisia	373	*3.9*	394	*4.2*	-21	*3.4*	386	*3.1*	381	*3.5*	5	*3.1*
Uruguay	418	*4.2*	439	*2.8*	-21	*3.9*	429	*4.0*	418	*3.1*	11	*4.0*

See notes at end of table.

Table C-11. Continued

Jurisdiction	Using scientific evidence					
	Male		Female		Male-Female Difference	
	Average	s.e.	Average	s.e.	Average	s.e.
OECD average	**498**	***0.8***	**501**	***0.7***	**-3**	***0.8***
OECD jurisdictions						
Australia	530	*3.4*	533	*3.0*	-3	*4.2*
Austria	509	*4.9*	500	*6.2*	9	*6.1*
Belgium	512	*3.8*	521	*3.8*	-9	*4.7*
Canada	541	*2.7*	542	*2.3*	-1	*2.3*
Czech Republic	501	*5.0*	500	*5.4*	1	*6.5*
Denmark	490	*4.1*	487	*4.0*	3	*3.8*
Finland	564	*3.0*	571	*2.7*	-7	*3.3*
France	509	*5.0*	513	*4.2*	-4	*4.7*
Germany	517	*5.6*	513	*4.5*	4	*4.3*
Greece	456	*5.6*	475	*3.7*	-20	*5.4*
Hungary	497	*4.1*	498	*4.5*	-1	*5.2*
Iceland	487	*3.1*	495	*2.5*	-7	*4.4*
Ireland	503	*4.8*	509	*3.5*	-7	*4.8*
Italy	466	*3.2*	468	*3.1*	-2	*4.2*
Japan	543	*5.8*	545	*6.4*	-2	*8.9*
Korea, Republic of	535	*5.2*	542	*4.5*	-8	*6.4*
Luxembourg	493	*2.0*	490	*2.2*	3	*3.5*
Mexico	404	*3.7*	401	*3.0*	3	*2.7*
Netherlands	527	*3.8*	524	*3.7*	3	*3.5*
New Zealand	532	*4.4*	541	*4.3*	-10	*5.8*
Norway	469	*4.2*	476	*3.9*	-7	*3.8*
Poland	492	*3.0*	495	*3.0*	-3	*2.8*
Portugal	473	*4.2*	471	*4.0*	2	*3.8*
Slovak Republic	478	*4.8*	478	*3.6*	#	*t*
Spain	484	*3.4*	485	*3.1*	-1	*2.5*
Sweden	494	*3.1*	499	*3.2*	-5	*3.4*
Switzerland	520	*3.6*	517	*3.9*	2	*2.9*
Turkey	410	*5.2*	426	*4.6*	-16	*4.7*
United Kingdom	517	*3.1*	510	*3.1*	6	*3.8*
United States	486	*6.1*	491	*4.6*	-5	*4.1*

Table C-11. Continued

Jurisdiction	Using scientific evidence					
	Male		Female		Male-Female Difference	
	Average	s.e.	Average	s.e.	Average	s.e.
OECD average	**498**	***0.8***	**501**	***0.7***	**-3**	***0.8***
Non-OECD jurisdictions						
Argentina	374	*7.4*	396	*7.7*	-23	*6.2*
Azerbaijan	342	*4.5*	347	*3.9*	-6	*2.4*
Brazil	382	*3.9*	375	*3.8*	6	*2.7*
Bulgaria	404	*8.0*	430	*8.2*	-26	*6.7*
Chile	447	*6.2*	431	*5.2*	16	*5.3*
Chinese Taipei	532	*4.5*	532	*5.1*	#	*†*
Colombia	386	*4.5*	381	*4.8*	5	*4.9*
Croatia	488	*4.1*	493	*3.5*	-5	*4.8*
Estonia	529	*3.2*	533	*3.0*	-5	*3.3*
Hong Kong-China	544	*3.8*	541	*4.0*	2	*5.5*
Indonesia	388	*10.2*	383	*5.0*	5	*7.3*
Israel	456	*6.7*	464	*5.4*	-8	*7.6*
Jordan	385	*5.5*	424	*3.6*	-39	*6.3*
Kyrgyz Republic	280	*4.7*	295	*3.9*	-15	*3.7*
Latvia	484	*4.1*	497	*3.6*	-13	*3.6*
Liechtenstein	524	*8.2*	544	*6.8*	-20	*12.2*
Lithuania	478	*3.7*	495	*3.3*	-17	*3.0*
Macao-China	512	*2.0*	511	*1.6*	#	*†*
Qatar	307	*1.5*	341	*1.9*	-35	*2.5*
Republic of Montenegro	403	*2.0*	411	*2.0*	-8	*3.1*
Republic of Serbia	419	*4.0*	431	*4.8*	-11	*4.9*
Romania	403	*6.0*	412	*6.7*	-9	*4.6*
Russian Federation	478	*4.5*	483	*4.4*	-5	*3.1*
Slovenia	510	*2.3*	522	*2.0*	-12	*3.4*
Thailand	409	*4.2*	433	*2.7*	-24	*4.5*
Tunisia	377	*4.1*	387	*4.3*	-10	*3.9*
Uruguay	425	*4.0*	433	*3.5*	-8	*4.1*

† Not applicable. # Rounds to zero.

Note: The Organization for Economic Cooperation and Development (OECD) average is the average of the national averages of the OECD member jurisdictions. Because the Program for International Student Assessment (PISA) is principally an OECD study, the results for non-OECD jurisdictions are displayed separately from those of the OECD jurisdictions and are not included in the OECD average. Differences were computed using unrounded numbers. Because of an error in printing the test booklets, the United States mean performance may be misestimated by approximately 1 score point. The impact is below one standard error. For details see appendix B. Standard error is noted by *s.e.*

Source: Organization for Economic Cooperation and Development (OECD), Program for International Student Assessment (PISA), 2006.

Table C-12. Average scores of U.S. 15-year-old students on combined science literacy scale, by race/ethnicity: 2006

Race/ethnicity	Average	*s.e.*
U.S. average	**489**	***4.2***
White, non-Hispanic	523	*3.0*
Black, non-Hispanic	409	*8.8*
Hispanic	439	*4.7*
Asian, non-Hispanic	499	*9.7*
American Indian/Alaska Native, non-Hispanic	436	*12.0*
Native Hawaiian/Other Pacific Islander, non-Hispanic	483	*24.5*
More than one race, non-Hispanic	501	*8.0*
OECD average	**500**	***0.5***

Note: Black includes African American, and Hispanic includes Latino. Students who identified themselves as being of Hispanic origin were classified as Hispanic, regardless of their race. To reach a particular proficiency level, a student must correctly answer a majority of items at that level. Students were classified into science literacy levels according to their scores. Exact cut point scores are as follows: below level 1 (a score less than or equal to 334.94); level 1 (a score greater than 334.94 and less than or equal to 409.54); level 2 (a score greater than 409.54 and less than or equal to 484.14); level 3 (a score greater than 484.14 and less than or equal to 558.73); level 4 (a score greater than 558.73 and less than or equal to 633.33); level 5 (a score greater than 633.33 and less than or equal to 707.93); and level 6 (a score greater than 707.93). The Organization for Economic Cooperation and Development (OECD) average is the average of the national averages of the OECD member jurisdictions. Because of an error in printing the test booklets, the United States mean performance may be misestimated by approximately 1 score point. The impact is below one standard error. For details see appendix B. Standard error is noted by *s.e.*

Source: Organization for Economic Cooperation and Development (OECD), Program for International Student Assessment (PISA), 2006.

Appendix D: PISA 2006 Expert Panelists

Rodger Bybee
Executive Director
Biological Sciences Curriculum Study Colorado Springs, CO

John Easton
Executive Director
Consortium on Chicago School Research Chicago, IL

Thomas Hoffer
Senior Research Scientist
National Opinion Research Center Chicago, IL

Stan Metzenberg
Associate Professor, Department of Biology California State University at Northridge Northridge, CA
Brett Moulding

State Science Specialist
Utah State Office of Education Salt Lake City, UT

Aaron Pallas
Professor of Sociology and Education Columbia University
New York, NY

Jo Ellen Roseman
Director
American Association for the Advancement of Science
Washington, DC

Gerald Wheeler
Executive Director
National Science Teachers Association Arlington, VA

In: Mathematics and Science Education
Editor: Chad P. Allerton

ISBN: 978-1-60692-313-9

Chapter 3

THE NATION'S REPORT CARD MATHEMATICS 2007: NATIONAL ASSESSMENT OF EDUCATIONAL PROGRESS AT GRADES 4 AND 8

U.S. Department of Education

ABSTRACT

The Nation's Report Card™ informs the public about the academic achievement of elementary and secondary students in the United States. Report cards communicate the findings of the National Assessment of Educational Progress (NAEP), a continuing and nationally representative measure of achievement in various subjects over time.

For over three decades, NAEP assessments have been conducted periodically in reading, mathematics, science, writing, U.S. history, civics, geography, and other subjects. By collecting and reporting information on student performance at the national, state, and local levels, NAEP is an integral part of our nation's evaluation of the condition and progress of education. Only information related to academic achievement and relevant variables is collected.

The privacy of individual students and their families is protected, and the identities of participating schools are not released.

NAEP is a congressionally authorized project of the National Center for Education Statistics (NCES) within the Institute of Education Sciences of the U.S. Department of Education. The Commissioner of Education Statistics is responsible for carrying out the NAEP project. The National Assessment Governing Board oversees and sets policy for NAEP.

Executive Summary

Both fourth- and eighth-graders reached a higher level of performance in 2007 compared to earlier assessment years

The 2007 National Assessment of Educational Progress (NAEP) evaluated students' understanding of mathematics concepts and their ability to apply mathematics to everyday situations. Students demonstrated their knowledge of these critical skills by responding to questions about number properties and operations, measurement, geometry, data analysis and probability, and algebra.

A nationally representative sample of more than 350,000 students at grades 4 and 8 participated in the 2007 mathematics assessment. Comparing these results to results from previous years shows the progress fourth- and eighth-graders are making both in the nation and in individual states.

The average score for fourth-graders increased 27 points over the past 17 years, and the score for eighth-graders increased 19 points. Students at all levels of performance made gains, resulting in higher percentages of students at or above the *Basic* and *Prof cient* achievement levels.

Student groups make gains, few gaps narrow

As indicated in the chart on the following page, improvements for minority students did not always result in narrower achievement gaps with white students. White, black, and Hispanic students at both grades showed a better understanding of mathematics in 2007 when compared to all previous assessment years. However, when compared to the first assessment year in 1990, only the white – black score gap at grade 4 narrowed in 2007. The white – black score gap at grade 8 narrowed between 2005 and 2007.

Student groups	Grade 4 Since 1990	Grade 4 Since 2005	Grade 8 Since 1990	Grade 8 Since 2005
Overall	↑	↑	↑	↑
White	↑	↑	↑	↑
Black	↑	↑	↑	↑
Hispanic	↑	↑	↑	↑
Asian/Pacific Islander	↑	↑	↑	↔
American Indian/ Alaska Native	‡	↔	‡	↔
Gaps				
White – Black	↓	↔	↔	↓
White – Hispanic	↔	↔	↔	↔

↑ Indicates the score was higher or the gap increased in 2007.

↓ Indicates the score was lower or the gap decreased in 2007.

↔ Indicates there was no significant change in the score or the gap in 2007.

‡ Reporting standards not met. Sample size was insufficient to permit a reliable estimate.

The mathematics score for Asian/Pacific Islander students was higher in 2007 than in previous assessment years for grade 4, but at grade 8 showed no significant change from 2005 to 2007.

At both grades 4 and 8, scores rose for students regardless of their eligibility for the free and reduced-price school lunch program, a measure of socioeconomic status. Average scores were higher in 2007 than in 2005 for students who were eligible as well as for students who were not eligible.

EXAMPLES OF WHAT STUDENTS CAN DO IN MATHEMATICS

GRADE 4

80% identified a fraction modeled by a picture

64% determined the probability of a specific outcome

43% explained how to find the perimeter of a given shape

GRADE 8

71% estimated time given a rate and a distance

54% computed the measure of an angle in a figure

25% identified the graph of a linear equation

Fifteen States and Jurisdictions Make Gains at Both Grades

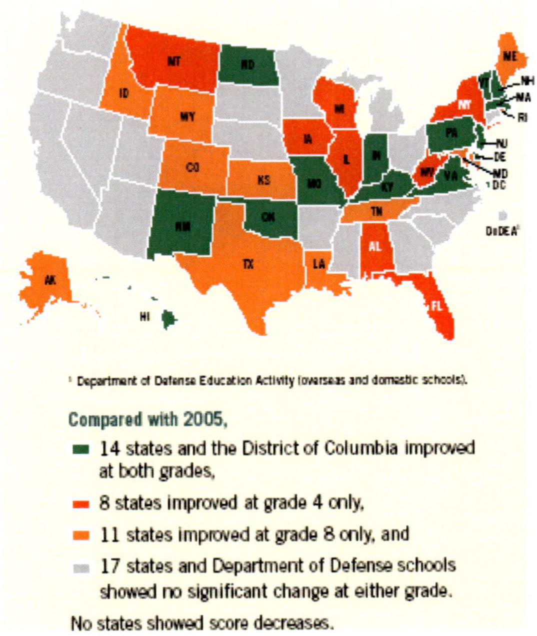
MT
ND
ID
WY
MI
NY
PA
IA
IL
IN
CO
KS
MO
KY
WV
VA
NM
OK
TN
AL
TX
LA
FL
AK
HI
ME
NH
MA
RI
NJ
DE
MD
DC
DoDEA[1]
[1] Department of Defense Education Activity (overseas and domestic schools).
Compared with 2005,
14 states and the District of Columbia improved at both grades,
8 states improved at grade 4 only,
11 states improved at grade 8 only, and
17 states and Department of Defense schools showed no significant change at either grade.
No states showed score decreases.

Differing patterns emerged when results were examined by different mathematics content areas. For example, 9 of the 29 states and jurisdictions that showed no change in overall performance at grade 4 did show a gain in at least one of the five content areas.

Overview of the Mathematics Assessment

With the belief that mathematics proficiency is integral to contemporary life, the NAEP mathematics assessment was designed to measure students' knowledge and skills in mathematics and their ability to apply their knowledge and skills in problem-solving situations.

The Mathematics Framework

The NAEP mathematics framework serves as the blueprint for the assessment, describing the specific mathematical skills that should be assessed at grades 4 and 8. Developed under the direction of the National Assessment Governing Board, the framework incorporates ideas and input from mathematicians, school administrators, policymakers, teachers, parents, and others.

MATHEMATICS CONTENT AREAS

Number properties and operations measures students' understanding of ways to represent, calculate, and estimate with numbers.

Measurement measures students' knowledge of measurement attributes, such as capacity and temperature, and geometric attributes, such as length, area, and volume.

Geometry measures students' knowledge and understanding of shapes in a plane and in space.

Data analysis and probability measures students' understanding of data representation, characteristics of data sets, experiments and samples, and probability.

Algebra measures students' understanding of patterns, using variables, algebraic representation, and functions.

The current NAEP mathematics framework was first used to guide the development of the 1990 assessment and has continued to be used through 2007. Updates to the framework over the years have provided more detail regarding the assessment design but did not change the content, allowing students' performance in 2007 to be compared with previous years. For more information on the framework, visit http://www.nagb.org/frameworks/math_07.pdf.

The framework details the mathematics objectives appropriate for grades 4 and 8. The topics covered by the framework include properties of numbers and operations, proportional reasoning, systems of measurement, relationships between geometric figures, data representation, probability, algebraic representations, equations and inequalities, and mathematical reasoning in various content areas.

Two dimensions of mathematics, *content areas* and *mathematical complexity*, are used to guide the assessment. Each item is designed to measure one of the five content areas. However, certain aspects of mathematics, such as computation, occur in all content areas. The level of complexity of a mathematics question is determined by the cognitive demands that it places on students.

Assessment Design

Because of the breadth of the content covered in the NAEP mathematics assessment, each student took just a portion of the test, consisting of two 25-minute sections. Testing time was divided evenly between multiple-choice and constructed-response (i.e., open-ended) questions. Some questions incorporated the use of rulers (at grade 4) or ruler/protractors (at grade 8), and some questions incorporated the use of geometric shapes or other manipulatives that were provided for students. On approximately one-third of the assessment, a four-function calculator was provided for students at grade 4, and a scientific calculator was provided for students at grade 8.

The distribution of items among each content area differs somewhat by grade to refl ect the knowledge and skills appropriate for each grade level. Table 1 shows the distribution across the content areas for grades 4 and 8, as recommended in the framework.

Table 1.Target percentage distribution of NAEP mathematics questions, by grade and content area: 2007

Content area	Grade 4	Grade 8
Number properties and operations	40%	20%
Measurement	20%	15%
Geometry	15%	20%
Data analysis and probability	10%	15%
Algebra	15%	30%

Source: U.S. Department of Education, National Assessment Governing Board, Mathematics Framework for the 2007 National Assessment of Educational Progress, 2006.

LEVELS OF MATHEMATICAL COMPLEXITY

Low complexity questions typically specify what a student is to do, which is often to carry out a routine mathematical procedure.

Moderate complexity questions involve more flexibility of thinking and often require a response with multiple steps.

High complexity questions make heavier demands and often require abstract reasoning or analysis in a novel situation.

REPORTING NAEP RESULTS

The students selected to take the NAEP assessment represent all fourth- and eighth-grade students across the U.S. Students who participate in NAEP play an important role by demonstrating the achievement of our nation's students and representing the success of our schooling. NAEP data can only be obtained with the cooperation of schools, teachers, and students nationwide.

Representative samples of schools and students at grades 4 and 8 participated in the 2007 NAEP mathematics assessment (table 2). The national results refl ect the performance of all fourth- and eighth-graders in public schools, private schools, Bureau of Indian Education schools, and Department of Defense schools. The state results refl ect the performance of students in public schools only.

Grade	Schools	Students
Grade 4	7,840	197, 7000
Grade 8	6,910	153,000

Note: The numbers of schools are rounded to the nearest ten, and the numbers of students are rounded to the nearest hundred.

Source: U.S. Department of Education, Institute of Education Sciences, National Center for Education Statistics, National Assessment of Educational Progress (NAEP), 2007 Mathematics Assessment.

National results from the 2007 mathematics assessment are compared to results from six previous assessment years for both grades 4 and 8. The 2007 state results are compared to results from five earlier assessments at grade 4 and six earlier assessments at grade 8. Changes in students' performance over time are summarized by comparing the results in 2007 to the next most recent assessment and the first assessment, except when pointing out consistent patterns across all assessments.

Scale Scores

NAEP mathematics results are reported on a 0–500 scale, overall and for each of the five content areas. Because NAEP scales are developed independently for each subject and for each content area within a subject, the scores cannot be compared across subjects or across content areas within the same subject. Results are also reported at five percentiles (10th, 25th, 50th, 75th, and 90th) to show trends in performance for lower-, middle-, and higher-performing students.

Achievement Levels

Based on recommendations from policymakers, educators, and members of the general public, the Governing Board sets specific achievement levels for each subject area and grade. Achievement levels are performance standards showing what students should know and be able to do. They provide another perspective with which to interpret student performance. NAEP results are reported as percentages of students performing at or above the *Basic* and *Proficient* levels and at the *Advanced* level.

NAEP ACHIEVEMENT LEVELS

Basic denotes partial mastery of prerequisite knowledge and skills that are fundamental for proficient work at a given grade.

Proficient represents solid academic performance. Students reaching this level have demonstrated competency over challenging subject matter.

Advanced represents superior performance.

As provided by law, NCES, upon review of congressionally mandated evaluations of NAEP, has determined that achievement levels are to be used on a trial basis and should be interpreted with caution. The NAEP achievement levels have been widely used by national and state officials.

Item Maps

Item maps provide another way to interpret the scale scores and achievement-level results for each grade. The item maps displayed in each grade section of this report show student performance on NAEP mathematics questions at different points on the scale.

Accommodations and Exclusions in NAEP

Testing accommodations, such as extra testing time or individual rather than group administration, are provided for students with disabilities or English language learners who could not fairly and accurately demonstrate their abilities without modified test administration procedures. Prior to 1996, no testing accommodations were provided in the NAEP mathematics assessment. This resulted in the exclusion of some students. In 1996, administration procedures were introduced at the national level allowing certain accommodations for students requiring such accommodations to participate. Accommodations for state level assessments began in 2000.

Note that most figures in this report show two data points in 1996—one permitting and the other not permitting accommodations. Both 1996 data points are presented in this report, but comparisons between 1996 and 2007 are based on accommodated samples.

Even with the availability of accommodations, there still remains a portion of students excluded from the NAEP assessment. Variations in exclusion and accommodation rates, due to differences in policies and practices regarding the identification and inclusion of students with disabilities and English language learners, should be considered when comparing students' performance over time and across states. While the effect of exclusion is not precisely known, comparisons of performance results could be affected if exclusion rates are comparatively high or vary widely over time. See appendix tables A-1 through A-5 for the percentages of students accommodated and excluded at the national and state levels. More information about NAEP's policy on inclusion of special-needs students is available at http://nces.ed.gov/nationsreportcard/about/inclusion.asp.

Interpreting Results

Changes in performance results over time may refl ect not only changes in students' knowledge and skills but also other factors, such as changes in student demographics, education programs and policies (including policies on accommodations and exclusions), and teacher qualifications.

NAEP results adopt widely accepted statistical standards; findings are reported based on a statistical significance level set at .05 with appropriate adjustments for multiple comparisons. In the tables and figures of this report that present results over time, the symbol (*) is used to indicate that a score or percentage in a previous assessment year is significantly different from the comparable measure in 2007. This symbol is also used in tables to highlight differences between male and female students within 2007. As a result of larger student sample sizes beginning in 2003, smaller differences (e.g., 1 or 2 points) can be found statistically significant than would have been detected with the smaller sample sizes used in earlier assessments.

Score differences or gaps cited in this report are calculated based on differences between unrounded numbers. Therefore, the reader may find that the score difference cited in the text may not be identical to the difference obtained from subtracting the rounded values shown in the accompanying tables or figures.

Not all of the data for results discussed in this report are presented in corresponding tables or figures. These and other results can be found in the NAEP Data Explorer at http://nces.ed.gov/nationsreportcard/nde.

For additional information, visit http://nationsreportcard.gov.

Scores Higher Than in All Previous Assessments

Results from the 2007 NAEP mathematics assessment revealed that fourth-graders' mathematical skills have improved over the last 17 years. Fourth-graders in 2007 scored 2 points higher than in 2005 and 27 points higher than in 1990 (figure 1).

Although not shown here, gains were also made in each of the mathematics content areas for which comparisons could be made back to 1990. Score point increases from 1990 to 2007 ranged from a 20-point gain in the measurement content area to a 30-point gain in algebra.

Figure 1. Trend in fourth-grade NAEP mathematics average scores

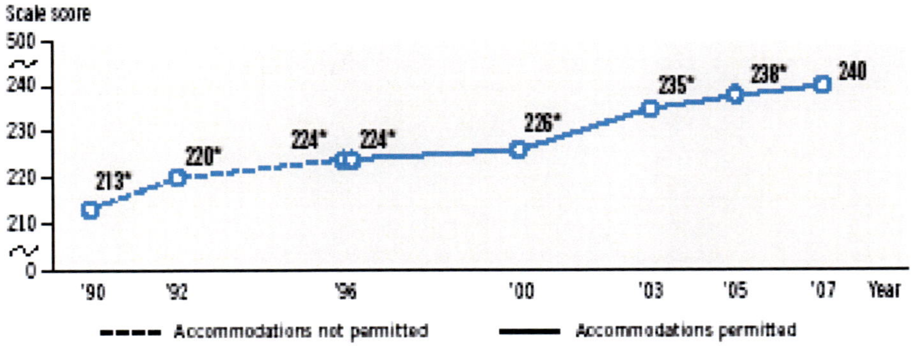

* Significantly different (p < .05) from 2007.
SOURCE: U.S. Department of Education, Institute of Education Sciences, National Center for Education Statistics, National Assessment of Educational Progress (NAEP), various years, 1990–2007 Mathematics Assessments.

Improvement Across All Performance Levels

The overall increase was seen at all levels of student performance. Lower-performing students (at the 10th and 25th percentiles), middle-performing students (at the 50th percentile), and higher-performing students (at the 75th and 90th percentiles) all scored higher in 2007 than in any previous assessment (figure 2). Lower-performing students made greater gains than higher-performing students over the last 17 years.

Figure 2. Trend in fourth-grade NAEP mathematics percentile scores

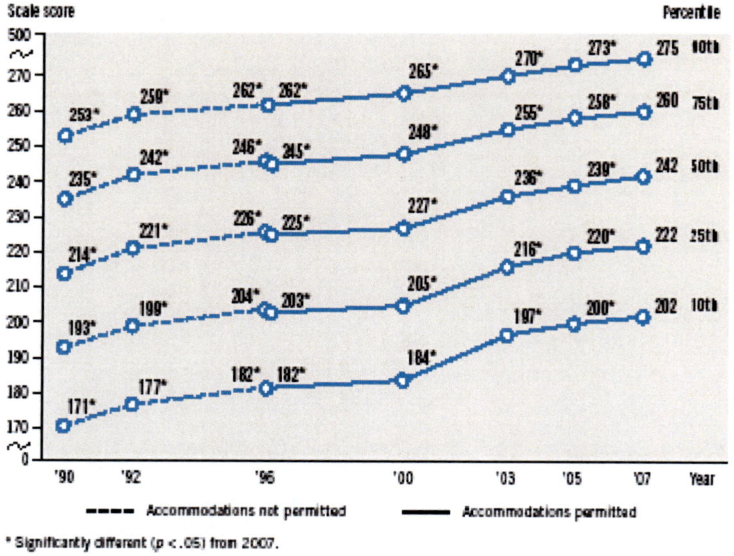

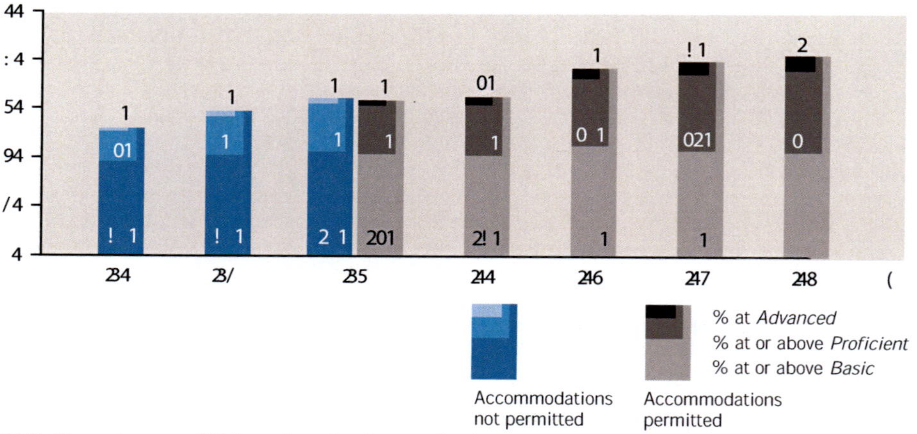

Source: U.S. Department of Education, Institute of Education Sciences, National Center for Education Statistics, National Assessment of Educational Progress (NAEP), various years, 1990–2007 Mathematics Assessments.

Figure 3. Trend in fourth-grade NAEP mathematics achievement-level performance

Score increases across all performance levels were also refl ected in the achievement-level results. The percentages of students at or above *Basic*, at or above *Proficient*, and at *Advanced* were higher in 2007 compared to the percentages for all previous assessment years (figure 3). The percentage of students at or above *Proficient* tripled from 13 percent in 1990 to 39 percent in 2007.

Most Racial/Ethnic Groups Show Gains

White, black, Hispanic, and Asian/Pacific Islander students all showed higher average mathematics scores in 2007 than in any of the previous assessments (figure 4). The 35-point[1] gain for black students from 1990 to 2007 was greater than the gains for white (28 points) and Hispanic students (27 points).

Figure 4. Trend in fourth-grade NAEP mathematics average scores, by race/ethnicity

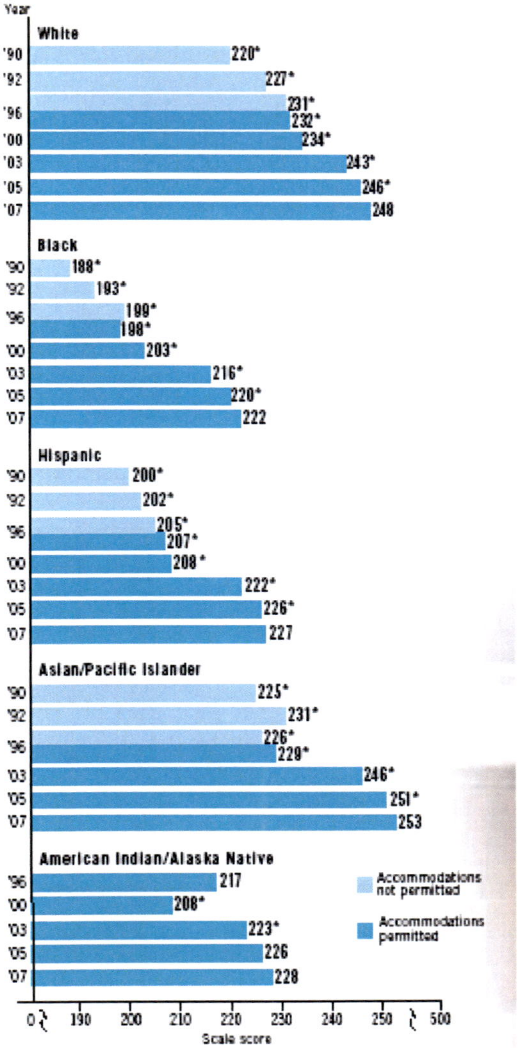

* Significantly different ($p < .05$) from 2007.

NOTE: Special analysis raised concerns about the accuracy and precision of national grade 4 Asian/Pacific Islander results in 2000. As a result, they are omitted from this figure. Sample sizes were insufficient to permit reliable estimates for American Indian/Alaska Native fourth-graders in 1990, 1992, and 1996 (accommodations not permitted sample). Black includes African American, Hispanic includes Latino, and Pacific Islander includes Native Hawaiian. Race categories exclude Hispanic origin.

SOURCE: U.S. Department of Education, Institute of Education Sciences, National Center for Education Statistics, National Assessment of Educational Progress (NAEP), various years, 1990–2007 Mathematics Assessments.

American Indian/Alaska Native students showed no significant score change since 2005. However, although not shown here, the percentage of this group of students performing at or above *Proficient* increased from 21 percent in 2005 to 25 percent in 2007.

Achievement-Level Results

Information is available on achievement-level results for racial/ethnic groups and other reporting categories athttp://nationsreportcard.gov/math_2007/data.asp.

Figure 5. Trend in fourth-grade NAEP mathematics average scores and score gaps, by selected racial/ethnic groups

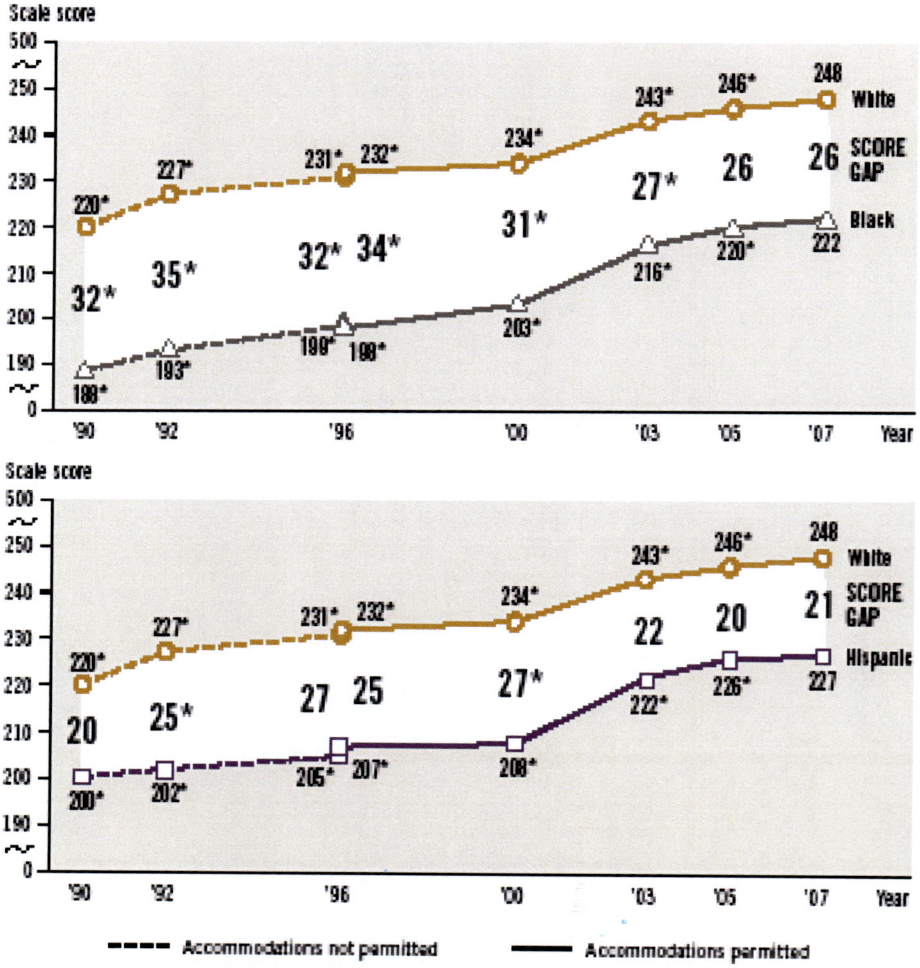

* Significantly different (p < .05) from 2007.
NOTE: Black includes African American, and Hispanic includes Latino. Race categories exclude Hispanic origin. Score gaps are calculated based on differences between unrounded average scores.

White–Black Gap Narrowing over Time

Score increases did not consistently result in a significant closing of performance gaps between minority students and white students. There was no significant change in the white–black score gap over the last two years (figure 5). Greater gains made by black students resulted in a smaller performance gap in 2007 compared to 17 years ago. The white–Hispanic gap was not significantly different from the gaps in either 2005 or 1990.

In each assessment year, NAEP collects information on student demographics. As shown in table 3, the percentage of white fourth-graders in the population was lower in 2007 than in previous assessment years, while the percentage of Hispanic students was higher. The percentage of Asian/Pacific Islander students was higher in 2007 than in 1990, and the percentage of black students was lower.

Table 3. Percentage of students assessed in fourth-grade NAEP mathematics, by race/ethnicity: Various years, 1990–2007

Race/ethnicity	1990	1992	1996	2000	2003	2005	2007
White	75*	73*	66*	64*	60*	58*	57
Black	18*	17*	16	16	17	16	16
Hispanic	6*	6*	11*	15*	18*	19*	20
Asian/Pacific Islander	1*	2*	5	—	4*	4	5
American Indian/ Alaska Native	1*	1	1	1	1	1	1

— Not available. Special analysis raised concerns about the accuracy and precision of national grade 4 Asian/Pacific Islander results in 2000. As a result, they are omitted from this table.

* Significantly different ($p < .05$) from 2007.

Note: Black includes African American, Hispanic includes Latino, and Pacific Islander includes Native Hawaiian. Race categories exclude Hispanic origin. Detail may not sum to totals because results are not shown for the "unclassified" race/ethnicity category.

Source: U.S. Department of Education, Institute of Education Sciences, National Center for Education Statistics, National Assessment of Educational Progress (NAEP), various years, 1990–2007 Mathematics Assessments.

Males score 2 points higher than females in 2007

Both male and female fourth graders showed improved mathematical skills, with higher scores in 2007 than in any of the previous assessment years (figure 6). Although both groups showed increases in 2007, male students scored 2 points higher on average than their female counterparts. The gap between the two groups in 2007 was not significantly different from the gaps in 2005 or 1990.

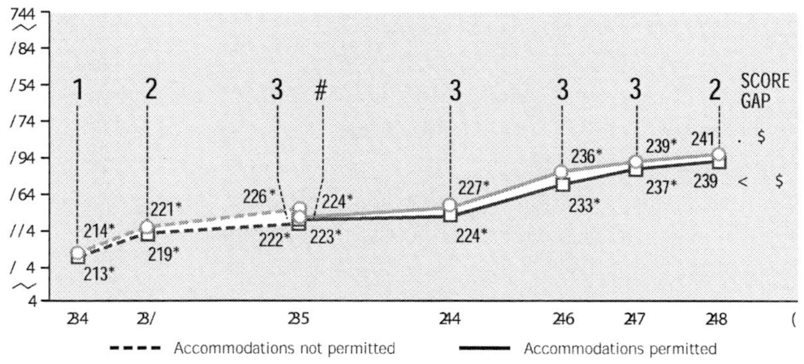

Rounds to zero.
* Significantly different ($p < .05$) from 2007.
Note: Score gaps are calculated based on differences between unrounded average scores.

Figure 6. Trend in fourth-grade NAEP mathematics average scores and score gaps, by gender

Differences in performance between male and female students in 2007 varied somewhat when examined by content area. Male students scored higher on average than female students in all the mathematics content areas with the exception of geometry in which female students scored higher (table 4)

Table 4. Average scores in fourth-grade NAEP mathematics, by content area and gender: 2007

Gender	Number properties and operations	Measurement	Geometry	Data analysis and probability	Algebra
Male	239*	241*	238*	244*	245*
Female	237	237	239	243	243

* Significantly different ($p < .05$) from female students in 2007.
Source: U.S. Department of Education, Institute of Education Sciences, National Center for Education Statistics, National Assessment of Educational Progress (NAEP), various years, 1990–2007 Mathematics Assessments.

Public School Students Score Lower Than Private School Students

Ninety-one percent of fourth-graders attended public schools in 2007, and 9 percent attended private schools. The average mathematics score for fourth-graders in public schools (239) was lower than for students in private schools overall (246) and in Catholic schools specifically (246).

Sample sizes for private schools as a whole were not always large enough to produce reliable estimates of student performance in some of the previous assessments, limiting the comparisons that can be made in performance over time (see the section on School and Student Participation Rates in the Technical Notes for more information). Trend results for public and Catholic school students, and for private school students in those years in which sample sizes were sufficient, are available at: http://nationsreportcard.gov/math_2007/m0038.asp.

Both Higher- and Lower-Income-Level Students Make Gains

A student's eligibility for free or reduced-price school lunch is used as an indicator of socioeconomic status; students from low-income families are typically eligible (eligibility criteria are described in the Technical Notes), while students from higher- income families typically are not.

Students who were not eligible continued to score higher on average than students who were eligible for free or reduced-price lunch; however, average mathematics scores were higher in 2007 than in 2005 for all three groups (figure 7). In 2007, those students eligible for reduced- price lunch had an average score 11 points higher than students eligible for free lunch.

Figure 7. Trend in fourth-grade NAEP mathematics average scores, by eligibility for free or reduced-price school lunch

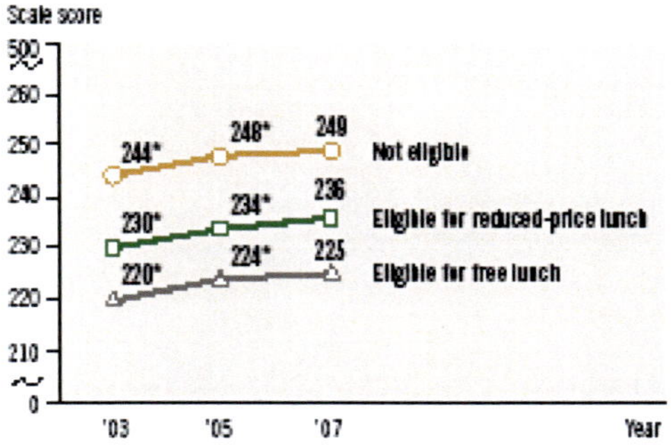

* Significantly different (p < .05) from 2007.

More than one-third of fourth graders assessed were eligible for free lunch in 2007 (table 5).

Changes in these percentages may refl ect not only a shift in the population but also changes in the National School Lunch Program and improvements in data quality. See the Technical Notes for more information.

State Performance at Grade 4

State results for public school students make it possible to compare each state's performance to other states and to the nation. All 50 states and 2 jurisdictions (i.e., the District of Columbia and Department of Defense schools) participated in the 2007 mathematics assessment. These 52 states and jurisdictions are all referred to as "states" in the following summary of state results. All states also participated in 2005, and 42 participated in the 1992 assessment, allowing for comparisons over time.

Table 5. Percentage of students assessed in fourth-grade NAEP mathematics, by eligibility for free or reduced-price school lunch: 2003, 2005, and 2007

Eligibility status	2003	2005	2007
Eligible for free lunch	33*	35	36
Eligible for reduced-price lunch	8*	7*	6
Not eligible	50*	50*	52
Information not available	10*	8*	7

Significantly different ($p < .05$) from 2007.

Note: Detail may not sum to totals because of rounding.

Source: U.S. Department of Education, Institute of Education Sciences, National Center for Education Statistics, National Assessment of Educational Progress (NAEP), 2003, 2005, and 2007 Mathematics Assessments.

Figure 8. Changes in fourth-grade NAEP mathematics average scores between 2005 and 2007

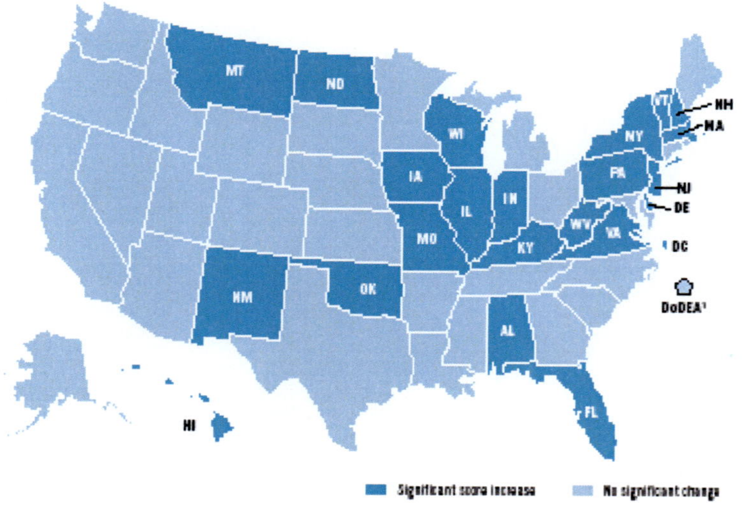

¹ Department of Defense Education Activity (overseas and domestic schools).
SOURCE: U.S. Department of Education, Institute of Education Sciences, National Center for Education Statistics National Assessment of Educational Progress (NAEP), 2005 and 2007 Mathematics Assessments.

Twenty-three States Show Score Increases

The map on the right highlights the 23 states in which overall average mathematics scores increased from 2005 to 2007 (figure 8). Of these 23 states, scores were also higher for white students in 14 states; black students in Delaware and New Jersey; Hispanic students in Delaware, Florida, Missouri, and New Mexico; Asian/Pacific Islander students in Hawaii; and American Indian/Alaska Native students in Oklahoma.

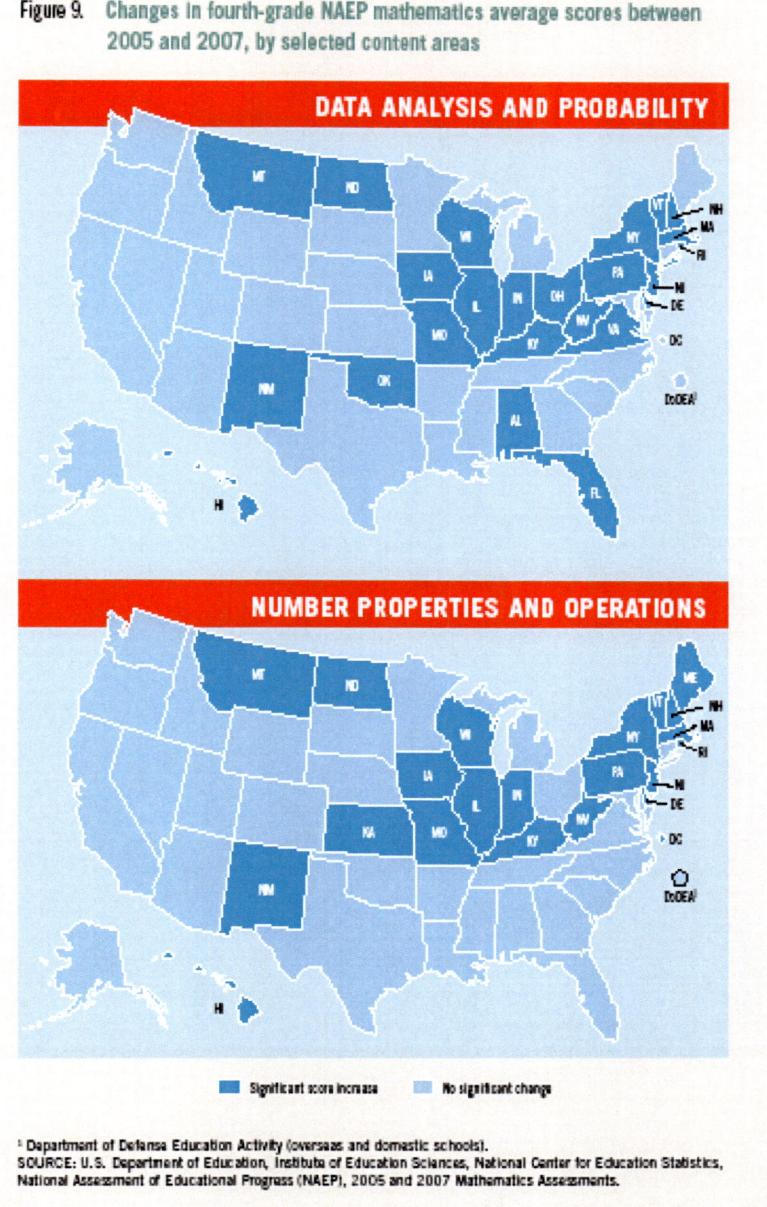

Figure 9. Changes in fourth-grade NAEP mathematics average scores between 2005 and 2007, by selected content areas

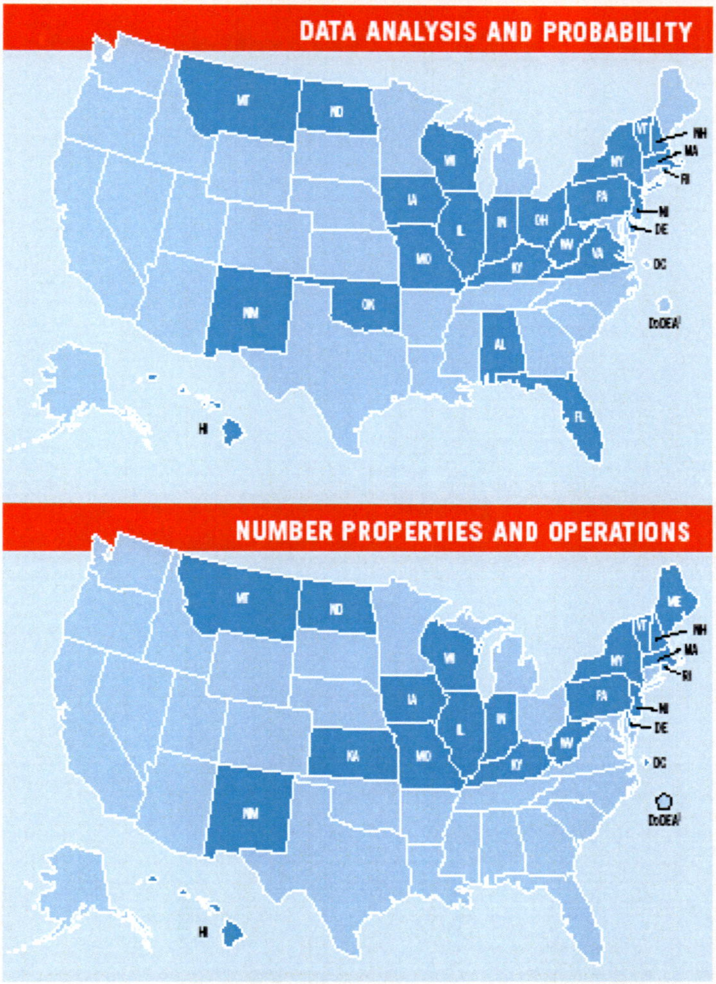

Significant score increase   No significant change

[1] Department of Defense Education Activity (overseas and domestic schools).
SOURCE: U.S. Department of Education, Institute of Education Sciences, National Center for Education Statistics, National Assessment of Educational Progress (NAEP), 2005 and 2007 Mathematics Assessments.

In no state did scores decline since 2005 for students overall or for any of the racial/ethnic groups. Scores increased since 1992 for all 42 states that participated in both 1992 and 2007. All of these states showed increases in the percentages of students both at or above *Basic* and at or above *Proficient*. These, and other state results for grade 4, are provided in figure 10, tables 6 and 7, and appendix tables A-7 through A-13.

Figure 10. Average scores and achievement-level results in NAEP mathematics for fourth-grade public school students, by state: 200

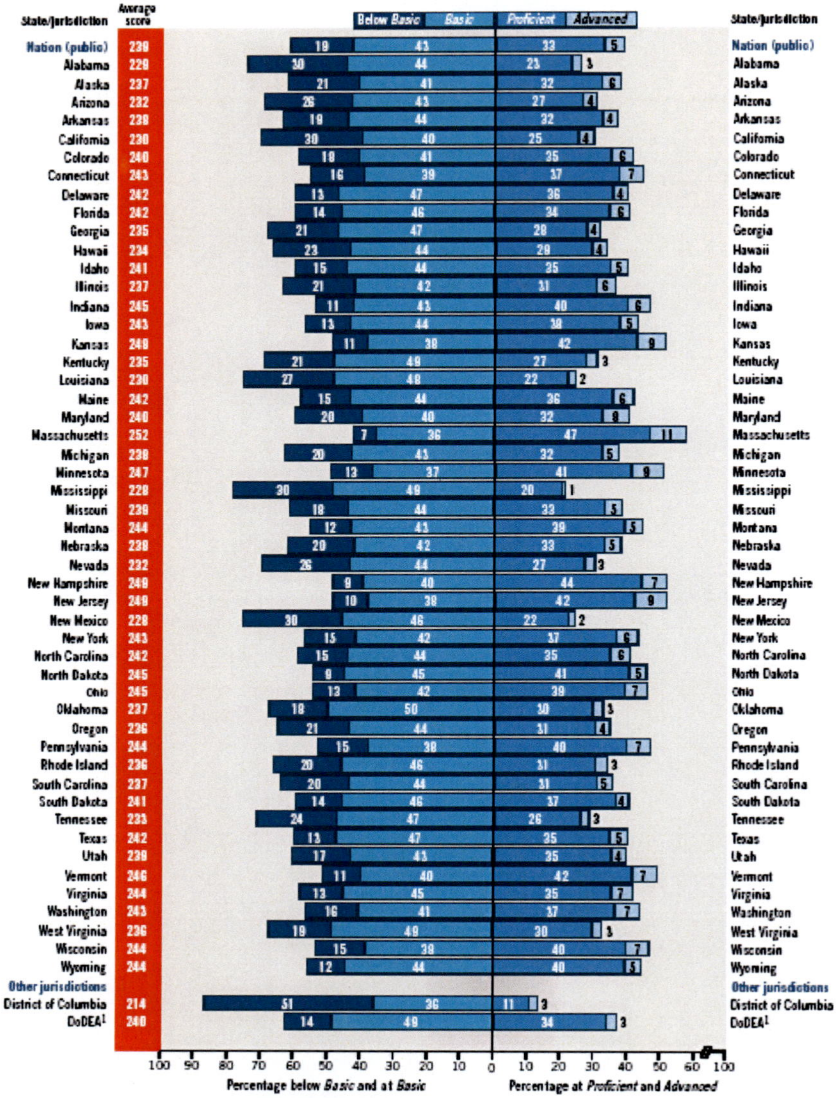

[1] Department of Defense Education Activity (overseas and domestic schools).
NOTE: The shaded bars are graphed using unrounded numbers. Detail may not sum to totals because of rounding.
SOURCE: U.S. Department of Education, Institute of Education Sciences, National Center for Education Statistics, National Assessment of Educational Progress (NAEP) Mathematics Assessment.

When making state comparisons, it is important to remember that performance results may be affected by differences in demographic makeup and exclusion and accommodation rates for students with disabilities and English language learners. Differences in performance could be affected if exclusion rates are comparatively high or vary widely over time. See appendix tables A-3 through A-5 for state exclusion and accommodation rates.

Table 6. Average scores in NAEP mathematics for fourth-grade public school students, by state: Various years, 1992–2007

State/jurisdiction	Accommodations not permitted			Accommodations permitted		
	1992	1996	2000	2000	2003	2005
Nation (public)[1]	**219***	**222***	**226***	**224***	**234***	**237***
Alabama	208*	212*	218*	217*	223*	225*
Alaska	—	224*	—	—	233*	236
Arizona	215*	218*	219*	219*	229*	230
Arkansas	210*	216*	217*	216*	229*	236
California	208*	209*	214*	213*	227*	230
Colorado	221*	226*	—	—	235*	239
Connecticut	227*	232*	234*	234*	241	242
Delaware	218*	215*	—	—	236*	240*
Florida	214*	216*	—	—	234*	239*
Georgia	216*	215*	220*	219*	230*	234
Hawaii	214*	215*	216*	216*	227*	230*
Idaho	222*	—	227*	224*	235*	242
Illinois	—	—	225*	223*	233*	233*
Indiana	221*	229*	234*	233*	238*	240*
Iowa	230*	229*	233*	231*	238*	240*
Kansas	—	—	232*	232*	242*	246
Kentucky	215*	220*	221*	219*	229*	231*
Louisiana	204*	209*	218*	218*	226*	230
Maine	232*	232*	231*	230*	238*	241
Maryland	217*	221*	222*	222*	233*	238
Massachusetts	227*	229*	235*	233*	242*	247*
Michigan	220*	226*	231*	229*	236	238
Minnesota	228*	232*	235*	234*	242*	246
Mississippi	202*	208*	211*	211*	223*	227
Missouri	222*	225*	229*	228*	235*	235*
Montana	—	228*	230*	228*	236*	241*
Nebraska	225*	228*	226*	225*	236	238
Nevada	—	218*	220*	220*	228*	230
New Hampshire	230*	—	—	—	243*	246*
New Jersey	227*	227*	—	—	239*	244*
New Mexico	213*	214*	214*	213*	223*	224*
New York	218*	223*	227*	225*	236*	238*
North Carolina	213*	224*	232*	230*	242	241
North Dakota	229*	231*	231*	230*	238*	243*
Ohio	219*	—	231*	230*	238*	242
Oklahoma	220*	—	225*	224*	229*	234*
Oregon	—	223*	227*	224*	236	238
Pennsylvania	224*	226*	—	—	236*	241*
Rhode Island	215*	220*	225*	224*	230*	233
South Carolina	212*	213*	220*	220*	236	238
South Dakota	—	—	—	—	237*	242
Tennessee	211*	219*	220*	220*	228*	232
Texas	218*	229*	233*	231*	237*	242
Utah	224*	227*	227*	227*	235*	239
Vermont	—	225*	232*	232*	242*	244*
Virginia	221*	223*	230*	230*	239*	240*
Washington	—	225*	—	—	238*	242
West Virginia	215*	223*	225*	223*	231*	231*
Wisconsin	229*	231*	—	—	237*	241*
Wyoming	225*	223*	229*	229*	241*	243
Other jurisdictions						
District of Columbia	193*	187*	193*	192*	205*	211*
DoDEA[2]	—	224*	228*	227*	237*	239

— Not available. The jurisdiction did not participate or did not meet the minimum participation guidelines for reporting.

* Significantly different ($p < .05$) from 2007 when only one jurisdiction or the nation is being examined.

[1] National results for assessments prior to 2003 are based on the national sample, not on aggregated state samples.

[2] Department of Defense Education Activity (overseas and domestic schools). Before 2005, DoDEA overseas and domestic schools were separate jurisdictions in NAEP. Pre-2005 data presented here were recalculated for comparability.

Note: State-level data were not collected in 1990.

Source: U.S. Department of Education, Institute of Education Sciences, National Center for Education Statistics, National Assessment of Educational Progress (NAEP), various years, 1992–2007 Mathematics Assessments.

Table 7. Percentage of fourth-grade public school students and average scores in NAEP mathematics, by selected student groups and state:

2007

	Race/ethnicity									
	White		Black		Hispanic		Asian/Pacifi c Islander		American Indian/ Alaska Native	
State/jurisdiction	Percentage of students	Average scale score	Percentage of students	Average scale score	Percentage of students	Average scale score	Percentage of students	Average scale score	Percentage of students	Av scale s
Nation (public)	**55**	**248**	**17**	**222**	**21**	**227**	**5**	**254**	**1**	
Alabama	58	238	37	213	3	218	1	‡	1	
Alaska	55	247	5	227	4	232	7	237	25	
Arizona	43	246	5	219	44	220	3	253	5	
Arkansas	67	245	22	217	9	230	2	236	1	
California	27	247	7	218	54	218	11	251	1	
Colorado	60	249	6	224	30	224	4	247	1	
Connecticut	64	252	13	220	18	223	5	255	#	
Delaware	54	249	33	230	10	234	3	261	#	
Florida	48	250	21	225	25	238	2	255	#	
Georgia	46	246	38	222	9	229	4	255	#	
Hawaii	17	244	3	230	4	224	63	233	1	
Idaho	81	245	1	‡	13	224	2	‡	3	
Illinois	56	248	19	216	19	223	4	257	#	
Indiana	78	249	10	224	7	233	1	‡	#	
Iowa	86	245	5	224	6	230	2	‡	#	
Kansas	73	252	8	226	13	234	2	260	1	
Kentucky	84	238	11	219	2	221	1	‡	#	
Louisiana	47	240	49	219	2	234	1	‡	1	
Maine	95	243	2	221	1	‡	2	‡	#	
Maryland	50	251	35	223	8	233	6	261	#	
Massachusetts	75	257	7	232	11	231	6	259	#	
Michigan	71	244	21	216	3	230	3	261	1	
Minnesota	78	252	8	222	7	229	5	239	2	
Mississippi	45	239	52	217	2	‡	1	‡	#	
Missouri	77	245	19	218	3	234	1	‡	#	
Montana	83	247	1	‡	3	241	1	‡	12	
Nebraska	75	244	7	211	14	220	1	‡	2	
Nevada	43	243	8	219	40	221	7	242	1	
New Hampshire	91	250	2	226	4	232	3	258	#	
New Jersey	57	255	14	232	20	234	8	267	#	
New Mexico	29	242	3	220	58	222	2	‡	9	
New York	53	251	19	225	20	230	8	260	#	
North Carolina	55	251	28	224	10	235	2	253	1	
North Dakota	87	248	2	‡	2	‡	1	‡	9	
Ohio	75	250	18	225	3	231	2	‡	#	
Oklahoma	58	242	11	220	9	227	2	247	20	
Oregon	71	241	3	219	17	217	5	249	2	
Pennsylvania	77	249	14	222	6	229	3	259	#	
Rhode Island	70	242	8	219	19	220	3	244	1	
South Carolina	57	248	36	221	4	227	1	‡	#	
South Dakota	83	245	2	221	2	228	1	‡	12	
Tennessee	69	240	26	214	3	222	1	‡	#	
Texas	36	253	15	230	45	236	3	263	#	
Utah	80	244	1	‡	15	220	2	244	2	
Vermont	94	247	2	‡	1	‡	2	‡	1	
Virginia	58	251	26	228	8	235	5	256	#	
Washington	65	248	6	222	15	225	11	250	2	
West Virginia	93	237	5	223	1	‡	1	‡	#	
Wisconsin	77	250	10	212	8	229	3	245	1	
Wyoming	84	246	2	‡	10	229	1	‡	3	
Other jurisdictions										
District of Columbia	6	262	84	209	9	220	2	‡	#	
DoDEA[1]	51	246	17	227	14	233	7	239	1	

See notes at end of table.

Table 7. Continued

2007

State/jurisdiction	Race/ethnicity									
	White		Black		Hispanic		Asian/Pacific Islander		American Indian/ Alaska Native	
	Percentage of students	Average scale score	Percentage of students	Average scale score	Percentage of students	Average scale score	Percentage of students	Average scale score	Percentage of students	Av scale s
Nation (public)	**55**	**248**	**17**	**222**	**21**	**227**	**5**	**254**	**1**	
Alabama	58	238	37	213	3	218	1	‡	1	
Alaska	55	247	5	227	4	232	7	237	25	
Arizona	43	246	5	219	44	220	3	253	5	
Arkansas	67	245	22	217	9	230	2	236	1	
California	27	247	7	218	54	218	11	251	1	
Colorado	60	249	6	224	30	224	4	247	1	
Connecticut	64	252	13	220	18	223	5	255	#	
Delaware	54	249	33	230	10	234	3	261	#	
Florida	48	250	21	225	25	238	2	255	#	
Georgia	46	246	38	222	9	229	4	255	#	
Hawaii	17	244	3	230	4	224	63	233	1	
Idaho	81	245	1	‡	13	224	2	‡	3	
Illinois	56	248	19	216	19	223	4	257	#	
Indiana	78	249	10	224	7	233	1	‡	#	
Iowa	86	245	5	224	6	230	2	‡	#	
Kansas	73	252	8	226	13	234	2	260	1	
Kentucky	84	238	11	219	2	221	1	‡	#	
Louisiana	47	240	49	219	2	234	1	‡	1	
Maine	95	243	2	221	1	‡	2	‡	#	
Maryland	50	251	35	223	8	233	6	261	#	
Massachusetts	75	257	7	232	11	231	6	259	#	
Michigan	71	244	21	216	3	230	3	261	1	
Minnesota	78	252	8	222	7	229	5	239	2	
Mississippi	45	239	52	217	2	‡	1	‡	#	
Missouri	77	245	19	218	3	234	1	‡	#	
Montana	83	247	1	‡	3	241	1	‡	12	
Nebraska	75	244	7	211	14	220	1	‡	2	
Nevada	43	243	8	219	40	221	7	242	1	
New Hampshire	91	250	2	226	4	232	3	258	#	
New Jersey	57	255	14	232	20	234	8	267	#	
New Mexico	29	242	3	220	58	222	2	‡	9	
New York	53	251	19	225	20	230	8	260	#	
North Carolina	55	251	28	224	10	235	2	253	1	
North Dakota	87	248	2	‡	2	‡	1	‡	9	
Ohio	75	250	18	225	3	231	2	‡	#	
Oklahoma	58	242	11	220	9	227	2	247	20	
Oregon	71	241	3	219	17	217	5	249	2	
Pennsylvania	77	249	14	222	6	229	3	259	#	
Rhode Island	70	242	8	219	19	220	3	244	1	
South Carolina	57	248	36	221	4	227	1	‡	#	
South Dakota	83	245	2	221	2	228	1	‡	12	
Tennessee	69	240	26	214	3	222	1	‡	#	
Texas	36	253	15	230	45	236	3	263	#	
Utah	80	244	1	‡	15	220	2	244	2	
Vermont	94	247	2	‡	1	‡	2	‡	1	
Virginia	58	251	26	228	8	235	5	256	#	
Washington	65	248	6	222	15	225	11	250	2	
West Virginia	93	237	5	223	1	‡	1	‡	#	
Wisconsin	77	250	10	212	8	229	3	245	1	
Wyoming	84	246	2	‡	10	229	1	‡	3	
Other jurisdictions										
District of Columbia	6	262	84	209	9	220	2	‡	#	
DoDEA[1]	51	246	17	227	14	233	7	239	1	

See notes at end of table.

* Significantly different ($p < .05$) from 2007 when only one jurisdiction or the nation is being examined.

[1] National results for assessments prior to 2003 are based on the national sample, not on aggregated state samples.

[2] Department of Defense Education Activity (overseas and domestic schools). Before 2005, DoDEA overseas and domestic schools were separate jurisdictions in NAEP. Pre-2005 data presented here were recalculated for comparability.

Note: State-level data were not collected in 1990.

Source: U.S. Department of Education, Institute of Education Sciences, National Center for Education Statistics, National Assessment of Educational Progress (NAEP), various years, 1992–2007 Mathematics Assessments.

States' Progress Varies by Mathematics Content Areas

While scores for the mathematics content areas cannot be directly compared to one another, examining patterns in differences over time shows that changes in overall results for a state may not always be consistent with changes for any particular content area. Among the 23 states posting overall gains between 2005 and 2007, 6 states—Indiana, Kentucky, Massachusetts, Missouri, New York, and West Virginia—scored higher in all five of the mathematics content areas. Among the 29 states with no overall change, Kansas, Maine, Maryland, Nevada, Ohio, Texas, and the Department of Defense schools showed increases in one content area; Rhode Island and Wyoming increased in two content areas; and Oregon decreased in two content areas. The two maps presented on the right show changes from 2005 to 2007 in states' scores for two of the five mathematics content areas: data analysis and probability and number properties and operations (figure 9). The data analysis and probability content area had the most score increases, with 24 states making gains. In the number properties and operations content area, which accounts for the largest percentage of assessment questions, 22 states showed increases.

For More Information...

State Comparison Tool orders states by students' performance overall and for student groups both within an assessment year and based on changes across years (http://nces.ed.gov/nationsreportcard/nde/statecom p).

State Profiles provide information on each state's school and student population and a summary of its NAEP results (http://nces.ed.gov/ nationsreportcard/states).

Assessment Content at Grade 4

To interpret the results in meaningful ways, it is important to understand the content of the assessment. Content was varied to reflect differences in the skills students were expected to have at each grade. The proportion of the assessment devoted to each of the mathematics content areas in each grade can be found in the overview section of this report.

Of the 166 questions that made up the fourth-grade mathematics assessment, the largest percentage (40 percent) focused on number properties and operations. It was expected that fourth-graders should have a solid grasp of whole numbers and a beginning understanding of fractions.

In measurement, the emphasis was on length, including perimeter, distance, and height. Students were expected to demonstrate knowledge of common customary and metric units. In geometry, students were expected to be familiar with simple figures in 2- and 3-dimensions and their attributes. In data analysis and probability, students were expected to demonstrate understanding of how data are collected and organized and basic concepts of probability. In algebra at this grade, the emphasis was on recognizing, describing, and extending patterns and rules.

Mathematics Achievement Levels at Grade 4

The following descriptions are abbreviated versions of the full achievement-level descriptions for grade 4 mathematics. The cut score depicting the lowest score representative of that level is noted in parentheses.

Basic (214): Fourth-graders performing at the *Basic* level should be able to estimate and use basic facts to perform simple computations with whole numbers; show some understanding of fractions and decimals; and solve some simple real-world problems in all NAEP content areas. Students at this level should be able to use—though not always accurately—four-function calculators, rulers, and geometric shapes. Their written responses are often minimal and presented without supporting information.

Proficient (249): Fourth-graders performing at the *Prof cient* level should be able to use whole numbers to estimate, compute, and determine whether results are reasonable. They should have a conceptual understanding of fractions and decimals; be able to solve real-world problems in all NAEP content areas; and use four-function calculators, rulers, and geometric shapes appropriately. Students performing at the *Prof cient* level should employ problem-solving strategies such as identifying and using appropriate information. Their written solutions should be organized and presented both with supporting information and explanations of how they were achieved.

Advanced (282): Fourth-graders performing at the *Advanced* level should be able to solve complex nonroutine real-world problems in all NAEP content areas. They should display mastery in the use of four-function calculators, rulers, and geometric shapes. These students are expected to draw logical conclusions and justify answers and solution processes by explaining why, as well as how, they were achieved. They should go beyond the obvious in their interpretations and be able to communicate their thoughts clearly and concisely.

The full descriptions can be found at http://www.nagb.org/frameworks/math_07.pdf.

What Fourth-Graders Know and Can Do in Mathematics

The item map below is useful for understanding performance at different levels on the scale. The scale scores on the left represent the average scores for students who were likely to get the items correct. The lower-boundary scores at each achievement level are noted in boxes. The descriptions of selected assessment questions are listed on the right along with the corresponding mathematics content areas.

For example, the map on this page shows that fourth- graders performing in the middle of the *Basic* range (students with an average score of 225) were likely to be able to identify a fraction modeled by a picture. Students performing in the middle of the *Proficient* range (with an average score of 267) were likely to be able to explain how to find the perimeter of a given shape.

GRADE 4 NAEP MATHEMATICS ITEM MAP

	Scale score	Content area	Question description
	500		
	~		
Advanced	330	Data analysis and probability	Label sections in a spinner to satisfy a given condition
Advanced	318	Number properties and operations	*Add three fractions with like denominators*
Advanced	296	Algebra	*Relate input to output from a table of values*
Advanced	294	Number properties and operations	*Solve a story problem involving addition and subtraction* (shown on page 22)
Advanced	290	Measurement	*Find area of a square with inscribed triangle*
Advanced	289	Geometry	*Recognize the result of folding a given shape*
Advanced	287	Data analysis and probability	Identify color with highest chance of being chosen (shown on page 23)
	282		
Proficient	279	Number properties and operations	Solve a story problem requiring multiple operations
Proficient	279	Data analysis and probability	*Identify picture representing greatest probability*
Proficient	267	Measurement	Explain how to find the perimeter of a given shape
Proficient	264	Number properties and operations	*Solve a story problem involving money*
Proficient	263	Algebra	*Identify number that would be in a pattern*
Proficient	262	Geometry	*Determine the number of blocks used to build a figure*
Proficient	255	Number properties and operations	*Use place value to determine the amount of increase*
Proficient	250	Geometry	*Identify the 3-D shape resulting from folding paper*
Proficient	249	Data analysis and probability	*Determine probability of a specific outcome*
	249		
Basic	245	Number properties and operations	Recognize property of odd numbers
Basic	243	Number properties and operations	Multiply two decimal numbers
Basic	232	Measurement	*Determine attribute being measured from a picture*
Basic	230	Number properties and operations	*Subtract a three-digit number from a four-digit number*
Basic	227	Algebra	*Identify number sentence that models a balanced scale*
Basic	225	Number properties and operations	*Identify a fraction modeled by a picture*
Basic	220	Algebra	*Identify an expression that represents a scenario*
Basic	218	Number properties and operations	*Find a sum based on place value*
Basic	217	Geometry	*Identify congruent triangles*
	214		
	211	Data analysis and probability	Complete a bar graph
	205	Geometry	*Use reason to identify figure based on description*
	202	Measurement	*Identify appropriate unit for measuring length*
	202	Number properties and operations	*Identify place value representation of a number*
	191	Algebra	*Find unknown in whole number sentence*
	~		
	0		

NOTE: Regular type denotes a constructed-response question. *Italic* type denotes a multiple-choice question. The position of a question on the scale represents the average scale score attained by students who had a 65 percent probability of successfully answering a constructed-response question, or a 74 percent probability of correctly answering a four-option multiple-choice question. For constructed-response questions, the question description represents students' performance rated as completely correct. Scale score ranges for mathematics achievement levels are referenced on the map.

SOURCE: U.S. Department of Education, Institute of Education Sciences, National Center for Education Statistics, National Assessment of Educational Progress (NAEP), 2007 Mathematics Assessment.

Sample Question about Number Properties and Operations

This sample question measures fourth-graders' performance in the number properties and operations content area. In particular, it addresses the "Number operations" subtopic, which focuses on computation, the effects of operations on numbers, and the relationships between operations. The framework objective measured is "Solve application problems involving numbers and operations." Students were not permitted to use a calculator to solve this problem.

Thirty-six percent of fourth-graders selected the correct answer (choice B). One way to arrive at this answer is to first use subtraction to determine that the bridge was built in 1926, and then use addition to determine that it was 50 years old in 1976. The most common incorrect answer (choice A), which was selected by 39 percent of fourth- graders, can be obtained by subtracting 50 years from 2001. The other incorrect answer choices (C and D) represent computation errors.

Percentage of fourth-grade students in each response category in 2007

Choice A	Choice B	Choice C	Choice D	Omitted
39	36	10	14	1

Note: Detail may not sum to totals because of rounding

The table below shows the percentage of fourth-graders within each achievement level who answered this question correctly. For example, 27 percent of fourth-graders at the *Basic* level selected the correct answer choice.

Percentage correct for fourth-grade students at each achievement level in 2007

Overall	**Below**	**Basic At**	**Basic At**	**Profi cient At**	**Advanced**
36	**24**	**27**	**46**	**77**	

Source: U.S. Department of Education, Institute of Education Sciences, National Center for Education Statistics, National Assessment of Educational Progress (NAEP), 2007 Mathematics Assessment.

Sample Question about Data Analysis and Probability

This sample question measures fourth-graders' performance in the data analysis and probability content area. It addresses the "Probability" subtopic, which focuses on simple probability and counting or representing the outcomes of a given event. The framework objective measured by this question is "Use informal probabilistic thinking to describe chance events." Students were not permitted to use a calculator to solve this problem.

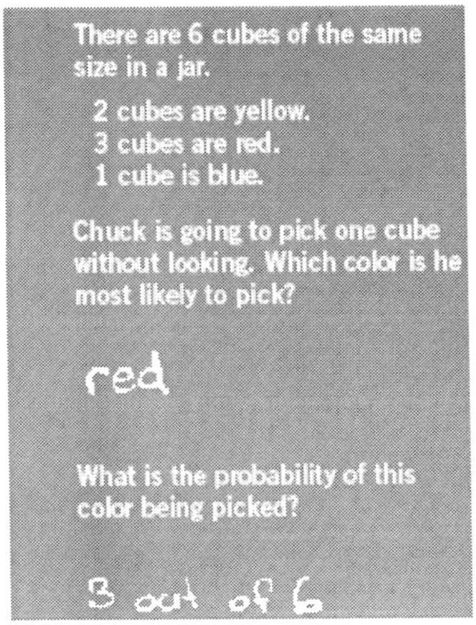

Student responses for this question were rated using the following three-level scoring guide:

Correct—Response indicates that a red cube is most likely to be picked and indicates that the probability is 3 out of 6 (or equivalent).

Partial—Response indicates that a red cube is most likely to be picked or indicates that the probability is 3 out of 6 (or equivalent).

Incorrect—All incorrect responses.

The student response on the right was rated as "Correct" because both parts of the question were answered correctly. Twenty-two percent of fourth-graders gave a response that was rated "Correct" for this question. Sixty-seven percent of fourth-graders provided a response rated as "Partial."

Percentage of fourth-grade students in each response category in 2007

Correct	Partial	Incorrect	Omitted
22	67	10	1

Progress (NAEP), 2007 Mathematics Assessment

8TH GRADE

Increased Mathematics Knowledge at Grade 8

Similar to the results for grade 4, the mathematical ability of eighth- graders also continued an upward trend in 2007. The average score in 2007 was higher than the score in any previous assessment. Students scored 3 points higher in 2007 than in 2005 and 19 points higher than in 1990[2] (figure 11).

Figure 11. Trend in eighth-grade NAEP mathematics average scores

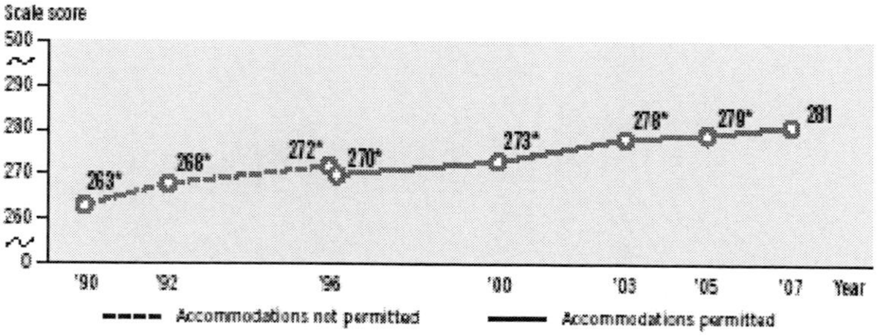

* Significantly different (p < .05) from 2007.
SOURCE: U.S. Department of Education, Institute of Education Sciences, National Center for Education Statistics, National Assessment of Educational Progress (NAEP), various years, 1990–2007 Mathematics Assessments.

Although not shown here, gains were also made in each of the five mathematics content areas. Score point increases from 1990 to 2007 ranged from a 13-point gain in number properties and operations to a 24-point gain in algebra.

Improvement at All Performance Levels

The improvement in mathematics at grade 8 was seen across all performance levels. Scores for students at each of the percentiles were higher in 2007 than the comparable scores from all previous years. Score increases since 1990 were almost even across the percentiles and ranged from 18 to 20 points (figure 12).

Figure 12. Trend in eighth-grade NAEP mathematics percentile scores

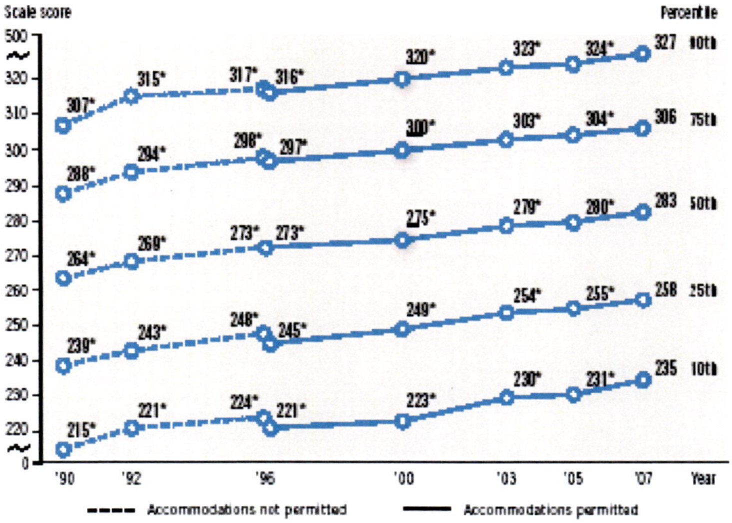

* Significantly different (p < .05) from 2007.

Figure 13. Trend in eighth-grade NAEP mathematics achievement-level performance

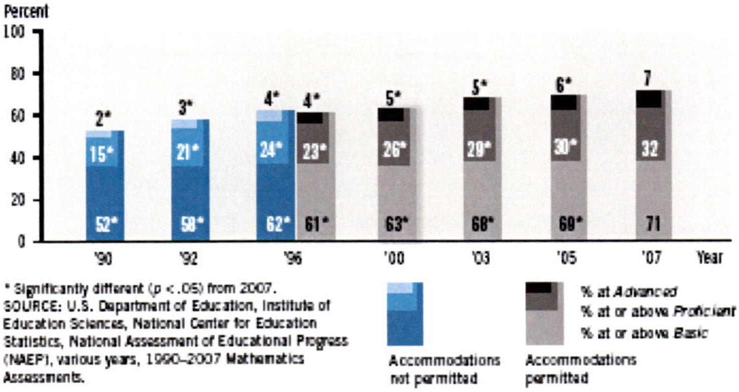

* Significantly different (p < .05) from 2007.
SOURCE: U.S. Department of Education, Institute of Education Sciences, National Center for Education Statistics, National Assessment of Educational Progress (NAEP), various years, 1990–2007 Mathematics Assessments.

Achievement-level results were consistent with the overall scale score and percentile results, showing improvement for students at all achievement levels. The percentages of students at or above *Basic*, at or above *Proficient*, and at *Advanced* were higher in 2007 than in all six previous assessment years (figure 13). The percentage of students at or above *Basic* increased 2 points since 2005 and 19 points in comparison to

1990. The percentage of students at or above *Proficient* doubled from 15 percent in 1990 to 32 percent in 2007, and the percentage at *Advanced* increased from 2 to 7 percent over the same period.

Figure 14. Trend in eighth-grade NAEP mathematics average scores, by race/ethnicity

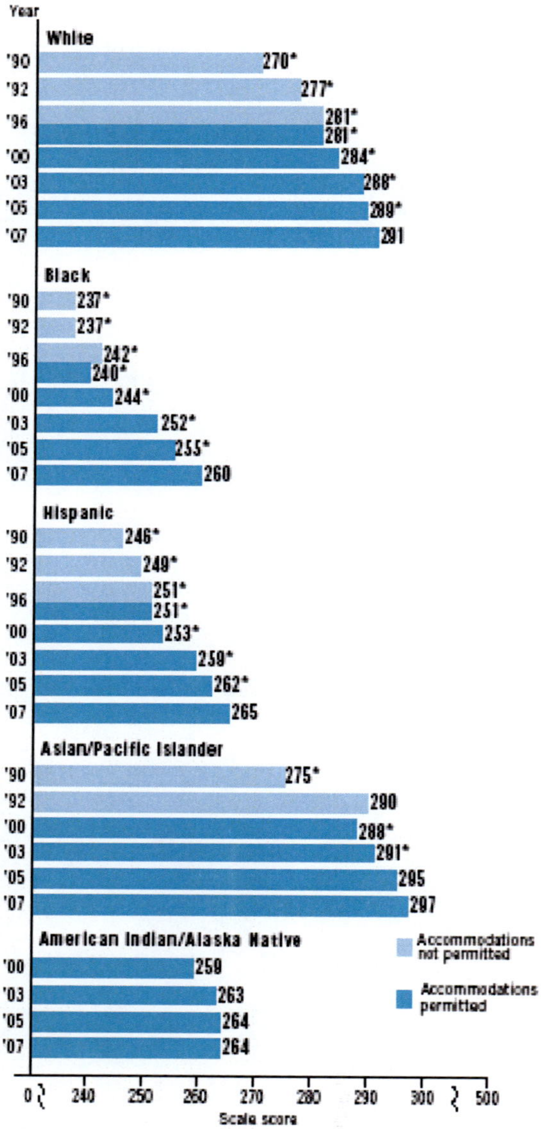

* Significantly different ($p < .05$) from 2007.
NOTE: Special analysis raised concerns about the accuracy and precision of national grade 8 Asian/Pacific Islander results in 1996. As a result, they are omitted from this figure. Sample sizes were insufficient to permit reliable estimates for American Indian/Alaska Native eighth-graders in 1990, 1992, and 1996. Black includes African American, Hispanic includes Latino, and Pacific Islander includes Native Hawaiian. Race categories exclude Hispanic origin.
SOURCE: U.S. Department of Education, Institute of Education Sciences, National Center for Education Statistics, National Assessment of Educational Progress (NAEP), various years, 1990–2007 Mathematics Assessments.

Gains for White, Black, and Hispanic students

The overall improved performance of eighth graders was not refl ected in all of the five student racial/ethnic groups. white, black, and Hispanic students showed higher average mathematics scores in 2007 than in all previous assessment years. The score for Asian/Pacific Islander students showed no significant change in comparison to 2005, but was higher than in 1990. No significant change in the score for American Indian/Alaska Native students was seen when compared to previous assessment years (figure 14).

White – Black gap narrows since 2005

Significant score gaps persisted between white students and their black and Hispanic peers. At 32 points, the white–black student score gap in 2007 was smaller than it was in 2005, but not significantly different from the gap in 1990.

Figure 15. Trend in eighth-grade NAEP mathematics average scores and score gaps, by selected racial/ethnic groups

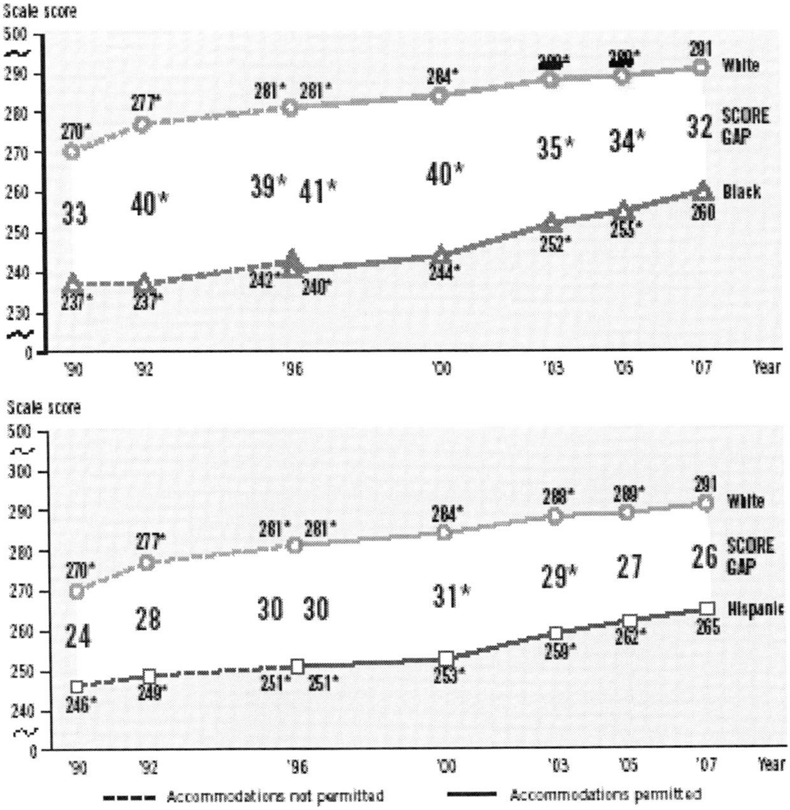

* Significantly different ($p < .05$) from 2007.

NOTE: Black includes African American, and Hispanic includes Latino. Race categories exclude Hispanic origin. Score gaps are calculated based on differences between unrounded average scores.

Table 8. Percentage of students assessed in eighth-grade NAEP mathematics, by race/ethnicity: Various years, 1990–2007

Race/ethnicity	1990	1992	1996	2000	2003	2005	2007
White	73*	73*	69*	65*	63*	61*	59
Black	16	16	17	16	16	16	16
Hispanic	7*	8*	10*	13*	15*	16*	18
Asian/ Pacific Islander	2*	2*	—	4	4	5	5
American Indian/ Alaska Native	1	1*	1	2	1	1	1

— Not available. Special analysis raised concerns about the accuracy and precision of national grade 8 Asian/Pacific Islander results in 1996. As a result, they are omitted from this table.

* Significantly different ($p < .05$) from 2007.

Note: Black includes African American, Hispanic includes Latino, and Pacific Islander includes Native Hawaiian. Race categories exclude Hispanic origin. Detail may not sum to totals because results are not shown for the "unclassified" race/ethnicity category.

Source: U.S. Department of Education, Institute of Education Sciences, National Center for Education Statistics, National Assessment of Educational Progress (NAEP), various years, 1990–2007 Mathematics Assessments.

The white–Hispanic score gap of 26 points was not significantly different from the gaps in either 2005 or 1990 (figure 15).

The percentage of white eighth graders in the population was lower in 2007 than in previous assessments, while the percentage of Hispanic students was higher (table 8). The percentage of Asian/Pacific Islander students in 2007 was not significantly different from 2005, but was higher than in 1990.

Both males and females make gains

As seen in grade 4, both male and female eighth graders showed improved mathematical performance. Higher scores were seen in 2007 than in any of the previous assessment years (figure 16).

In 2007, male students scored 2 points higher on average than their female counterparts. The gap between the two groups in 2007 was not statistically different from the gaps seen in 2005 and 1990.

As in grade 4, differences between male and female students varied somewhat when examined by content area in 2007. With the exception of geometry and data analysis and probability, male students scored higher on average than female students in the mathematics content areas (table 9). Female students scored 1 point higher in data analysis and probability. There was no significant difference in the performance of male and female students in geometry.

Figure 16. Trend in eighth-grade NAEP mathematics average scores and score gaps, by gender

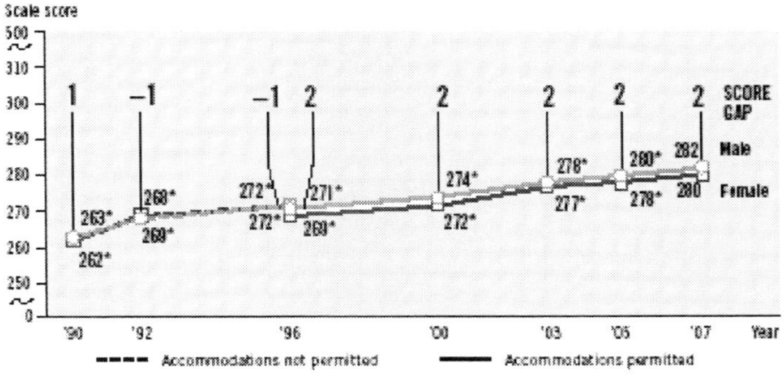

* Significantly different ($p < .05$) from 2007.
NOTE: Score gaps are calculated based on differences between unrounded average scores. Score gaps reflect the average scores for male students minus the scores for female students.

Gender	Number properties and operations	Measurement	Geometry	Data analysis and probability	Algebra
Male	282*	281 *	278	284*	286*
Female	277	275	278	285	285

* Significantly different ($p < .05$) from female students in 2007.

Source: U.S. Department of Education, Institute of Education Sciences, National Center for Education Statistics, National Assessment of Educational Progress (NAEP), various years, 1990–2007 Mathematics Assessments.

Gaps in Performance of Public and Private School Students

Ninety-one percent of eighth-graders attended public schools in 2007, and 9 percent attended private schools. The average mathematics score for eighth-graders in public schools (280) was lower than for students in private schools overall (293) and lower than for students in Catholic schools specifically (292).

Trend results for public and Catholic school students, and for private school students in those years in which sample sizes were sufficient, are available at: http://nationsreportcard.gov/math_2007/m0038.asp.

Improved Performance across Income Levels

Similar to the results for grade 4, scores increased for students who were eligible for either free or reduced-price school lunch as well as for students who were not eligible. Average mathematics scores were higher in 2007 than in 2005 for all three groups of students (figure 17).

Eighth-graders who were not eligible for free or reduced-price lunch scored higher on average than those who were eligible in 2007, and students eligible for reduced-price lunch scored higher than those eligible for free lunch.

Figure 17. Trend in eighth-grade NAEP mathematics average scores, by eligibility for free or reduced-price school lunch

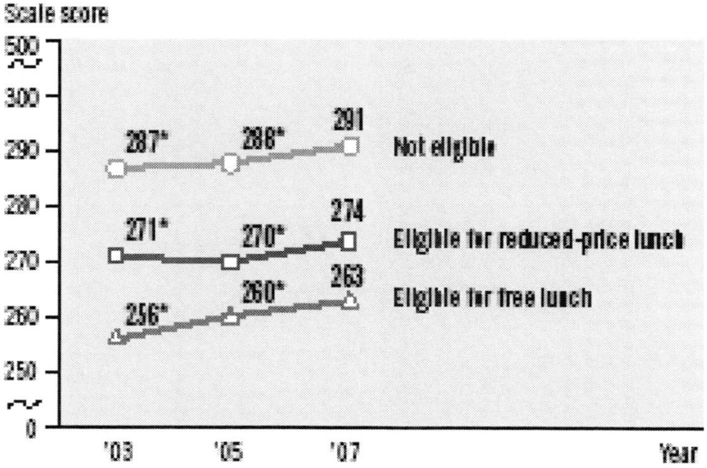

* Significantly different ($p < .05$) from 2007.

Changes over time in the percentages of students based on their eligibility for free or reduced-price school lunch are presented in table 10. About one-third of eighth graders assessed were eligible for free lunch in 2007.

Table 10. Percentage of students assessed in eighth-grade NAEP mathematics, by eligibility for free or reduced-price school lunch: 2003, 2005, and 2007

Eligibility status	2003	2005	2007
Eligible for free lunch	26*	29*	32
Eligible for reduced-price lunch	7*	7*	6
Not eligible	55	56	55
Information not available	11*	8	7

* Significantly different ($p < .05$) from 2007.

Note: Detail may not sum to totals because of rounding.

Source: U.S. Department of Education, Institute of Education Sciences, National Center for Education Statistics, National Assessment of Educational Progress (NAEP), 2003, 2005, and 2007 Mathematics Assessments.

STATE PERFORMANCE AT GRADE 8

All of the 52 states and jurisdictions that participated in 2007 also participated in 2005, and 38 participated in the 1990 assessment, allowing for comparisons over time. As with grade 4, it is important to remember that performance results for states may be affected by differences in demographic makeup and exclusion and accommodation rates for students with disabilities and English language learners, which may vary considerably across states as well as across years.

Increased scores in one-half of states

The map on the right highlights changes in states' average mathematics scores since 2005, with increases in 26 states (figure 18). Nine of these states showed increases for only students who were not eligible for free/reduced-price school lunch, while nine states showed increases for both students who were eligible and students who were not eligible.

Figure 18. Changes in eighth-grade NAEP mathematics average scores between 2005 and 2007

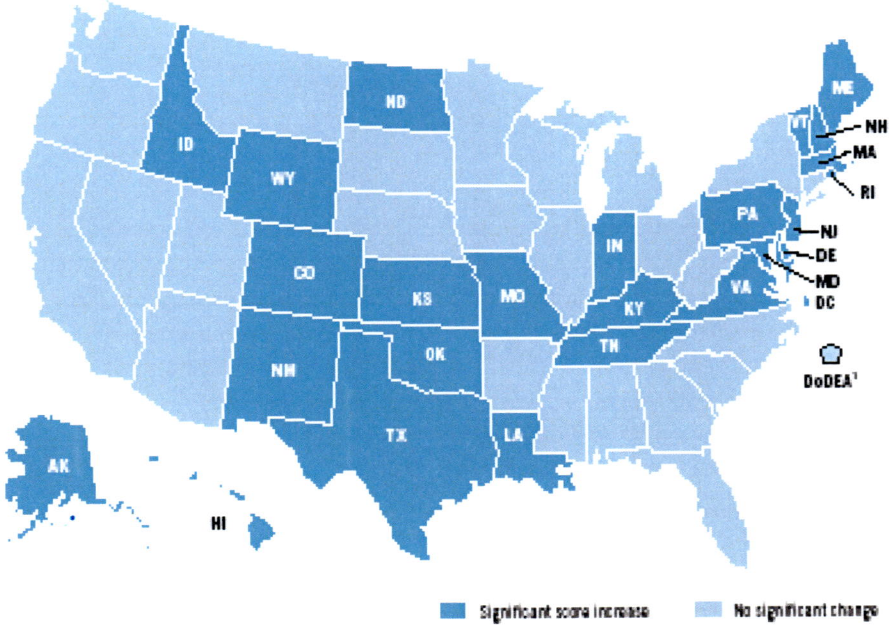

[1] Department of Defense Education Activity (overseas and domestic schools).
SOURCE: U.S. Department of Education, Institute of Education Sciences, National Center for Education Statistics, National Assessment of Educational Progress (NAEP), 2005 and 2007 Mathematics Assessments.

There were no states in which scores declined since 2005 for students overall.

All of the 38 states that participated in both 1990 and 2007 showed increases in average mathematics scores. These 38 states also showed increases in the percentages of students both at or above *Basic* and at or above *Proficient*. These and other state results for grade 8 are provided in figure 20, tables 11 and 12, and appendix tables A-14 through A-20.

Four states make gains in all content areas

Among the 26 states posting overall gains between 2005 and 2007, Kentucky, Massachusetts, Texas, and Wyoming were the only states that also scored higher in all five of the mathematics content areas.

Among the 26 states with no change in performance overall, 9 states (Arkansas, California, Florida, Minnesota, Mississippi, Nevada, North Carolina, Utah, and West Virginia) showed increases in one content area, Illinois increased in two content areas, and Montana increased in one area and decreased in another.

The two maps presented on the right show changes in states' average scores from 2005 to 2007 for two of the five mathematics content areas: algebra and measurement (figure 19).

The algebra and measurement content areas showed the most and fewest changes in state performance, respectively. Thirty states made gains in algebra, with no state posting a decline. The fewest states made gains in measurement, with increases in nine states and a decline in one state.

ASSESSMENT CONTENT AT GRADE 8

Of the 168 questions that made up the eighth-grade mathematics assessment, the largest percentage (approximately 30 percent) focused on algebra. The emphasis was on students' understanding of algebraic representations, patterns, and functions; linearity; and algebraic expressions, equations, and inequalities. The knowledge and skills expected at grade 8 in number properties and operations include computing with rational numbers, common irrational numbers, and numbers in scientific notation, and using numbers to solve problems involving proportionality and rates. In the measurement content area, students were expected to be familiar with area, volume, angles, and rates. In geometry, eighth- graders were expected to be familiar with parallel and perpendicular lines, angle relations in polygons, cross sections of solids, and the Pythagorean Theorem. In data analysis and probability, students were expected to use a variety of techniques for organizing and summarizing data, analyzing statistical claims, and demonstrating an understanding of the terminology and concepts of probability.

Mathematics Achievement Levels at Grade 8

The following descriptions are abbreviated versions of the full achievement-level descriptions for grade 8 mathematics. The cut score depicting the lowest score representative of that level is noted in parentheses.

Figure 19. Changes in eighth-grade NAEP mathematics average scores between 2005 and 2007, by selected content areas

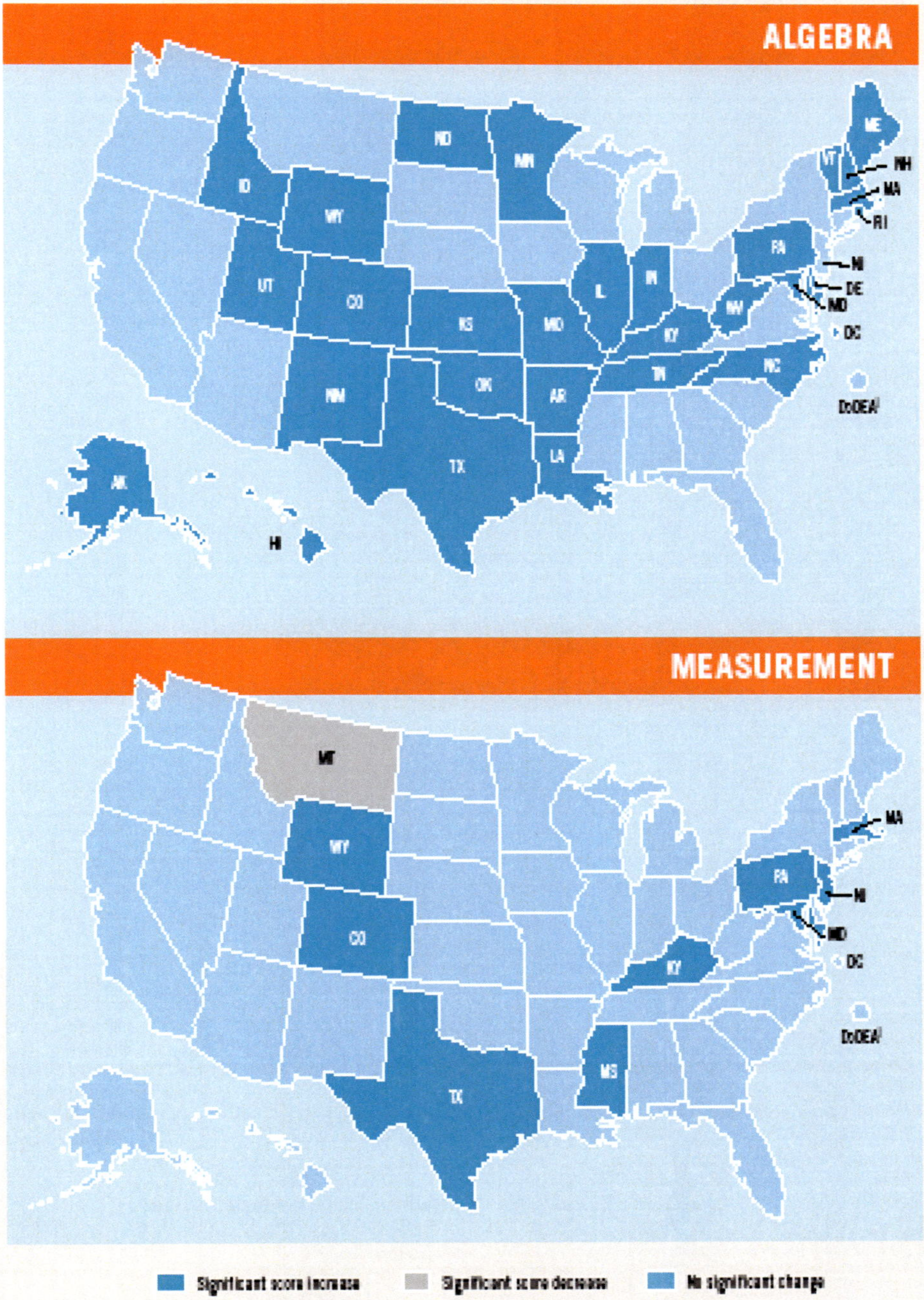

[1] Department of Defense Education Activity (overseas and domestic schools).
SOURCE: U.S. Department of Education, Institute of Education Sciences, National Center for Education Statistics, National Assessment of Educational Progress (NAEP), 2005 and 2007 Mathematics Assessments.

Figure 20. Average scores and achievement-level results in NAEP mathematics for eighth-grade public school students, by state: 2007

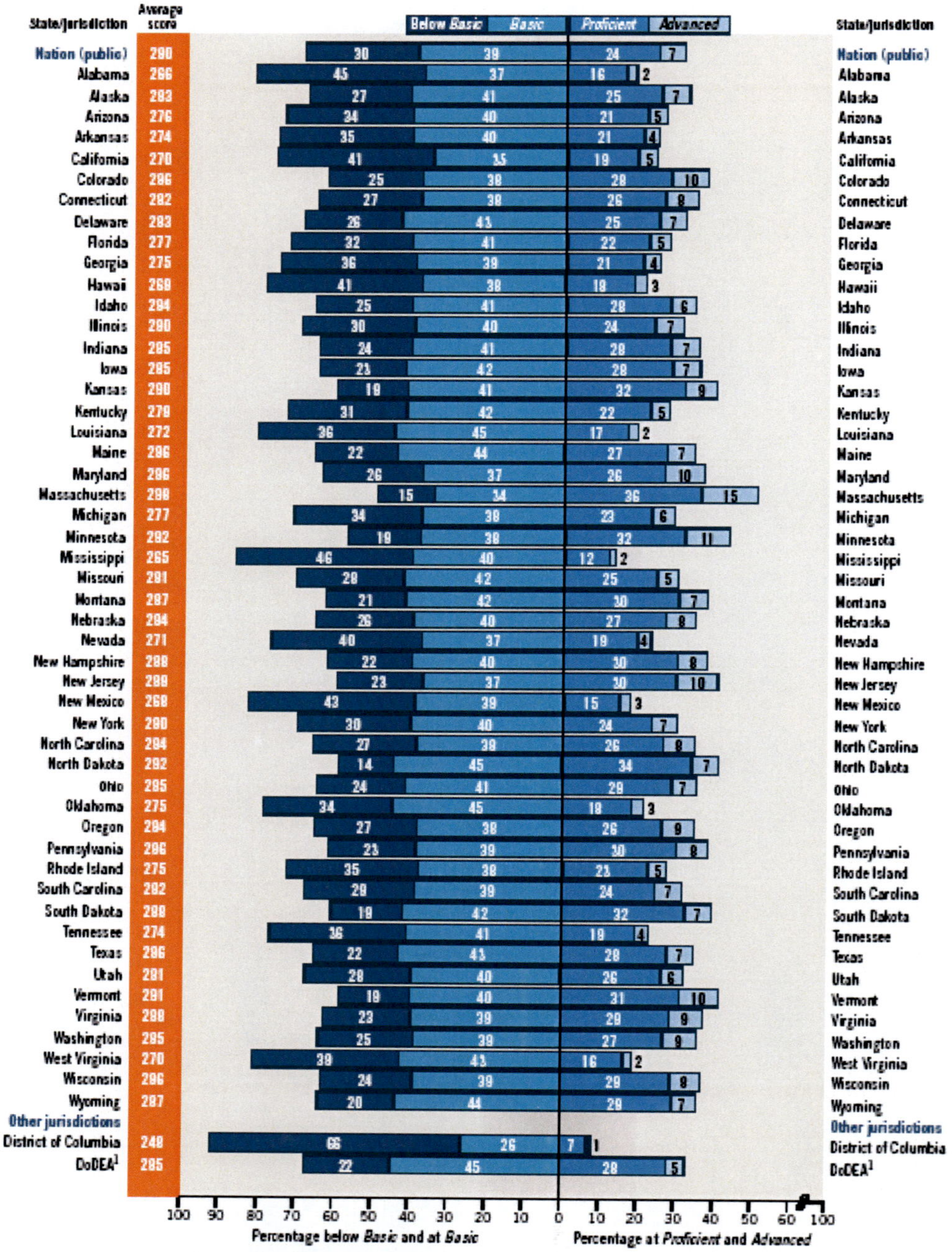

[1] Department of Defense Education Activity (overseas and domestic schools).

NOTE: The shaded bars are graphed using unrounded numbers. Detail may not sum to totals because of rounding.

SOURCE: U.S. Department of Education, Institute of Education Sciences, National Center for Education Statistics, National Assessment of Educational Progress (NAEP), 2007 Mathematics Assessment.

Table 11. Average scores in NAEP mathematics for eighth-grade public school students, by state: Various years, 1990–2007

	Accommodations not permitted				Accommodations permitted		
State/jurisdiction	1990	1992	1996	2000	2000	2003	2005
Nation (public)[1]	**262***	**267***	**271***	**274***	**272***	**276***	**278***
Alabama	253*	252*	257*	262	264	262	262
Alaska	—	—	278*	—	—	279*	279*
Arizona	260*	265*	268*	271*	269*	271*	274
Arkansas	256*	256*	262*	261*	257*	266*	272
California	256*	261*	263*	262*	260*	267*	269
Colorado	267*	272*	276*	—	—	283	281*
Connecticut	270*	274*	280	282	281	284	281
Delaware	261*	263*	267*	—	—	277*	281*
Florida	255*	260*	264*	—	—	271*	274
Georgia	259*	259*	262*	266*	265*	270*	272
Hawaii	251*	257*	262*	263*	262*	266*	266*
Idaho	271*	275*	—	278*	277*	280*	281*
Illinois	261*	—	—	277	275*	277*	278
Indiana	267*	270*	276*	283	281*	281*	282*
Iowa	278*	283	284	—	—	284	284
Kansas	—	—	—	284*	283*	284*	284*
Kentucky	257*	262*	267*	272*	270*	274*	274*
Louisiana	246*	250*	252*	259*	259*	266*	268*
Maine	—	279*	284	284*	281*	282*	281*
Maryland	261*	265*	270*	276*	272*	278*	278*
Massachusetts	—	273*	278*	283*	279*	287*	292*
Michigan	264*	267*	277	278	277	276	277
Minnesota	275*	282*	284*	288*	287*	291	290
Mississippi	—	246*	250*	254*	254*	261*	262
Missouri	—	271*	273*	274*	271*	279	276*
Montana	280*	—	283*	287	285	286	286
Nebraska	276*	278*	283	281*	280*	282	284
Nevada	—	—	—	268*	265*	268*	270
New Hampshire	273*	278*	—	—	—	286	285*
New Jersey	270*	272*	—	—	—	281*	284*
New Mexico	256*	260*	262*	260*	259*	263*	263*
New York	261*	266*	270*	276	271*	280	280
North Carolina	250*	258*	268*	280*	276*	281	282
North Dakota	281*	283*	284*	283*	282*	287*	287*
Ohio	264*	268*	—	283	281*	282	283
Oklahoma	263*	268*	—	272	270*	272	271*
Oregon	271*	—	276*	281	280	281	282
Pennsylvania	266*	271*	—	—	—	279*	281*
Rhode Island	260*	266*	269*	273	269*	272*	272*
South Carolina	—	261*	261*	266*	265*	277*	281
South Dakota	—	—	—	—	—	285*	287
Tennessee	—	259*	263*	263*	262*	268*	271*
Texas	258*	265*	270*	275*	273*	277*	281*
Utah	—	274*	277*	275*	274*	281	279
Vermont	—	—	279*	283*	281*	286*	287*
Virginia	264*	268*	270*	277*	275*	282*	284*
Washington	—	—	276*	—	—	281*	285
West Virginia	256*	259*	265*	271	266*	271	269
Wisconsin	274*	278*	283	—	—	284	285
Wyoming	272*	275*	275*	277*	276*	284*	282*
Other jurisdictions							
District of Columbia	231*	235*	233*	234*	235*	243*	245*
DoDEA[2]	—	—	274*	278*	277*	285	284

— Not available. The jurisdiction did not participate or did not meet the minimum participation guidelines for reporting.

* Significantly different ($p < .05$) from 2007 when only one jurisdiction or the nation is being examined.

[1] National results for assessments prior to 2003 are based on the national sample, not on aggregated state samples.

[2] Department of Defense Education Activity (overseas and domestic schools). Before 2005, DoDEA overseas and domestic schools were separate jurisdictions in NAEP. Pre-2005 data presented here were recalculated for comparability.

Source: U.S. Department of Education, Institute of Education Sciences, National Center for Education Statistics, National Assessment of Educational Progress (NAEP), various years, 1990–2007 Mathematics Assessments.

Table 12. Percentage of eighth-grade public school students and average scores in NAEP mathematics, by selected student groups and state:

2007

State/jurisdiction	Race/ethnicity									
	White		Black		Hispanic		Asian/Pacifi c Islander		American Indian/ Alaska Native	
	Percentage of students	Average scale score	Percentage of students	Average scale score	Percentage of students	Average scale score	Percentage of students	Average scale score	Percentage of students	Av scale
Nation (public)	**58**	**290**	**17**	**259**	**19**	**264**	**5**	**296**	**1**	
Alabama	60	278	35	246	2	249	1	‡	1	
Alaska	56	294	4	271	4	274	8	282	25	
Arizona	47	289	5	266	39	262	3	303	7	
Arkansas	69	282	22	254	7	256	1	‡	1	
California	31	287	7	253	48	256	12	293	1	
Colorado	65	296	7	272	25	264	3	297	1	
Connecticut	69	293	13	255	15	254	3	307	#	
Delaware	56	294	31	265	9	267	4	309	#	
Florida	48	289	23	259	24	270	2	293	#	
Georgia	46	288	43	261	7	266	2	‡	#	
Hawaii	14	278	2	‡	2	264	70	268	#	
Idaho	82	287	1	‡	14	264	1	‡	2	
Illinois	60	291	16	253	18	265	5	303	#	
Indiana	77	290	12	259	7	267	1	‡	#	
Iowa	88	288	4	257	6	261	2	‡	#	
Kansas	76	295	8	267	10	269	2	302	2	
Kentucky	86	282	10	257	2	‡	1	‡	#	
Louisiana	52	283	43	258	2	‡	2	‡	1	
Maine	96	287	2	‡	1	‡	1	‡	#	
Maryland	51	300	37	265	7	272	5	313	#	
Massachusetts	75	305	8	264	10	270	5	315	#	
Michigan	75	285	18	244	3	259	2	‡	1	
Minnesota	81	297	7	260	4	269	5	283	2	
Mississippi	47	279	51	251	1	‡	1	‡	#	
Missouri	75	288	19	253	3	270	2	‡	#	
Montana	85	291	1	‡	2	‡	1	‡	11	
Nebraska	80	291	7	240	11	261	1	‡	1	
Nevada	47	282	10	255	34	257	8	285	1	
New Hampshire	94	289	2	‡	3	264	1	‡	#	
New Jersey	57	298	17	264	19	271	7	314	#	
New Mexico	32	285	3	264	52	260	1	‡	12	
New York	55	290	19	258	18	264	6	302	1	
North Carolina	56	295	30	266	8	273	3	299	1	
North Dakota	89	295	1	‡	1	‡	1	‡	8	
Ohio	76	291	18	258	2	276	2	‡	#	
Oklahoma	59	280	9	258	8	259	2	‡	21	
Oregon	73	289	3	272	15	261	5	299	2	
Pennsylvania	76	293	15	257	6	264	3	314	#	
Rhode Island	70	284	9	250	17	251	4	282	1	
South Carolina	56	293	38	265	3	272	1	‡	#	
South Dakota	86	292	1	‡	2	269	1	‡	10	
Tennessee	67	282	28	254	4	264	2	‡	#	
Texas	38	300	15	271	44	277	3	309	#	
Utah	82	286	1	‡	12	256	3	277	2	
Vermont	95	292	1	‡	1	‡	2	‡	1	
Virginia	61	296	26	268	6	275	5	299	#	
Washington	69	291	5	264	14	263	10	289	2	
West Virginia	94	271	4	250	1	‡	1	‡	#	
Wisconsin	80	292	10	247	6	268	3	290	1	
Wyoming	86	290	1	‡	8	274	1	‡	3	
Other jurisdictions										
District of Columbia	3	‡	88	245	9	251	1	‡	#	
DoDEA[1]	48	291	18	272	15	282	8	284	1	

See notes at end of table.

Table 12. Continued

	Eligibility for free/reduced-price school lunch				Gender				
	Eligible		Not eligible		Male		Female		
State/jurisdiction	Percentage of students	Average scale score	Percentage of students	Average scale score	Percentage of students	Average scale score	Percentage of students	A\ scale	
Nation (public)	**41**	**265**	**58**	**291**	**51**	**281**	**49**		
Alabama	49	250	51	281	51	267	49		
Alaska	37	266	63	292	52	282	48		
Arizona	44	262	53	286	49	277	51		
Arkansas	51	263	49	285	48	274	52		
California	47	257	49	283	51	270	49		
Colorado	33	267	67	296	52	287	48		
Connecticut	27	256	73	292	51	282	49		
Delaware	33	270	67	290	51	285	49		
Florida	44	265	56	287	49	278	51		
Georgia	47	262	53	287	50	275	50		
Hawaii	42	258	58	276	52	267	48		
Idaho	39	273	60	290	49	285	51		
Illinois	39	262	61	292	50	282	50		
Indiana	36	271	64	293	52	286	48		
Iowa	30	270	70	292	51	287	49		
Kansas	36	275	64	299	50	291	50		
Kentucky	46	267	54	288	51	280	49		
Louisiana	57	264	42	284	48	273	52		
Maine	32	275	68	292	49	288	51		
Maryland	28	268	72	293	50	287	50		
Massachusetts	26	275	74	306	49	300	51		
Michigan	33	259	67	285	52	278	48		
Minnesota	26	273	72	298	51	292	49		
Mississippi	66	257	33	280	48	266	52		
Missouri	39	266	60	290	50	282	50		
Montana	34	272	65	295	50	287	50		
Nebraska	33	265	67	293	51	285	49		
Nevada	37	259	59	279	51	271	49		
New Hampshire	17	271	80	291	50	288	50		
New Jersey	27	266	71	297	51	290	49		
New Mexico	59	258	40	282	52	268	48		
New York	48	268	51	292	52	281	48		
North Carolina	44	268	55	296	50	285	50		
North Dakota	26	280	74	296	50	293	50		
Ohio	31	268	67	293	51	286	49		
Oklahoma	51	264	49	285	49	277	51		
Oregon	39	270	58	294	52	285	48		
Pennsylvania	29	267	71	294	51	289	49		
Rhode Island	33	257	67	285	52	276	48		
South Carolina	49	269	51	294	48	281	52		
South Dakota	30	275	70	294	52	290	48		
Tennessee	45	262	55	284	49	277	51		
Texas	50	275	50	297	50	287	50		
Utah	30	267	68	287	52	282	48		
Vermont	27	277	73	296	50	292	50		
Virginia	28	268	72	295	53	289	47		
Washington	33	268	65	294	50	285	50		
West Virginia	48	260	52	279	51	271	49		
Wisconsin	29	266	69	293	52	287	48		
Wyoming	28	275	72	291	52	288	48		
Other jurisdictions									
District of Columbia	65	243	35	259	46	248	54		
DoDEA[1]	#	‡	#	‡	49	285	51		

Rounds to zero.

‡ Reporting standards not met. Sample size is insufficient to permit a reliable estimate.

[1] Department of Defense Education Activity (overseas and domestic schools).

Note: Black includes African American, Hispanic includes Latino, and Pacific Islander includes Native Hawaiian. Race categories exclude Hispanic origin. Results are not shown for students whose race/ethnicity was "unclassified" and for students whose eligibility for free/reduced-price school lunch was not available.

Source: U.S. Department of Education, Institute of Education Sciences, National Center for Education Statistics, National Assessment of Educational Progress (NAEP), 2007 Mathematics Assessment.

GRADE 8 NAEP MATHEMATICS ITEM MAP

	Scale score	Content area	Question description
	500		
	~		
Advanced	364	**Geometry**	Model a geometrical situation given specific conditions
Advanced	355	**Measurement**	*Estimate side length of a square given area*
Advanced	342	**Algebra**	*Identify the graph of a linear equation*
Advanced	340	**Number properties and operations**	*Interpret a number expressed in scientific notation*
Advanced	337	**Geometry**	Find container height given dimensions of contents
Advanced	334	**Data analysis and probability**	Identify best method for selecting a sample
	333		
Proficient	329	**Algebra**	*Convert a temperature from Fahrenheit to Celsius*
Proficient	328	**Data analysis and probability**	*Identify which statistic is represented by a response*
Proficient	325	**Algebra**	Complete a table and write an algebraic expression
Proficient	320	**Number properties and operations**	*Determine distance given rate and time*
Proficient	317	**Number properties and operations**	Analyze a mathematical relationship (shown on page 39)
Proficient	314	**Algebra**	*Use a formula to solve a problem*
Proficient	311	**Number properties and operations**	*Divide large numbers in a given context*
Proficient	308	**Measurement**	Determine value of marks on a scale
Proficient	306	**Geometry**	*Determine measure of an angle in a figure*
Proficient	304	**Number properties and operations**	*Identify fractions listed in ascending order*
Proficient	301	**Algebra**	*Determine an equation relating sales and profit* (shown on page 38)
	299		
Basic	296	**Data analysis and probability**	*Identify relationship in a scatterplot*
Basic	296	**Number properties and operations**	*Convert raw points to a percentage*
Basic	287	**Data analysis and probability**	Explain which survey is better
Basic	278	**Number properties and operations**	*Estimate time given a rate and a distance*
Basic	276	**Algebra**	*Determine an expression to model a scenario*
Basic	268	**Measurement**	*Determine width after proportional enlargement*
Basic	265	**Algebra**	*Identify point on a graph with specified coordinates*
	262		
	261	**Algebra**	*Evaluate an expression for a specific value*
	259	**Data analysis and probability**	*Recognize misrepresented data*
	258	**Measurement**	*Determine dimensions that give the greatest volume*
	258	**Geometry**	*Identify the result of combining two shapes*
	257	**Algebra**	*Solve an algebraic equation*
	254	**Number properties and operations**	*Use place value to write a number*
	~		
	0		

NOTE: Regular type denotes a constructed-response question. *Italic* type denotes a multiple-choice question. The position of a question on the scale represents the average scale score attained by students who had a 65 percent probability of successfully answering a constructed-response question, a 74 percent probability of correctly answering a four-option multiple-choice question, or a 72 percent probability of correctly answering a five-option multiple-choice question. For constructed-response questions, the question description represents students' performance rated as completely correct. Scale score ranges for mathematics achievement levels are referenced on the map.
SOURCE: U.S. Department of Education, Institute of Education Sciences, National Center for Education Statistics, National Assessment of Educational Progress (NAEP), 2007 Mathematics Assessment.

Basic (262): Eighth-graders performing at the *Basic* level should complete problems correctly with the help of structural prompts such as diagrams, charts, and graphs. They should be able to solve problems in all NAEP content areas through the appropriate selection and use of strategies and technological tools, including calculators, computers, and geometric shapes. Students at this level also should be able to use fundamental algebraic and informal geometric concepts in problem solving. As they approach the *Proficient* level, students at the *Basic* level should be able to determine which of the available data are necessary and sufficient for correct solutions and use them in problem solving. However, these eighth graders show limited skill in communicating mathematically.

Proficient (299): Eighth-graders performing at the *Proficient* level should be able to conjecture, defend their ideas, and give supporting examples. They should understand the connections among fractions, percents, decimals, and other mathematical topics such as algebra and functions. Students at this level are expected to have a thorough understanding of *Basic* level arithmetic operations—an understanding sufficient for problem solving in practical situations. Quantity and spatial relationships in problem solving and reasoning should be familiar to them, and they should be able to convey underlying reasoning skills beyond the level of arithmetic. They should be able to compare and contrast mathematical ideas and generate their own examples. These students should make inferences from data and graphs, apply properties of informal geometry, and accurately use the tools of technology. Students at this level should understand the process of gathering and organizing data and be able to calculate, evaluate, and communicate results within the domain of statistics and probability.

Advanced (333): Eighth-graders performing at the *Advanced* level should be able to probe examples and counterexamples in order to shape generalizations from which they can develop models. Eighth-graders performing at the *Advanced* level should use number sense and geometric awareness to consider the reasonableness of an answer. They are expected to use abstract thinking to create unique problem-solving techniques and explain the reasoning processes underlying their conclusions.

What Eighth-Graders Know and Can Do in Mathematics

The item map below illustrates the range of mathematical knowledge and skills demonstrated by eighth-graders. For example, students performing near the middle of the *Basic* range (with an average score of 278) were likely to be able to estimate time given a rate and a distance. Students performing near the top of the *Proficient* range (with an average score of 325) were likely to be able to complete a table and write an algebraic expression.

Sample Question About Algebra

This sample question measures eighth-graders' performance in the algebra content area. It addresses the "Algebraic representations" subtopic, which focuses on analyzing, interpreting, and translating among different representations of linear relationships; representing points in a rectangular coordinate system; and recognizing common nonlinear relationships in meaningful contexts. The framework objective measured by this question is "Translate between different representations of linear expressions using symbols, graphs, tables, diagrams, or written descriptions." Students were permitted to use a calculator to solve this problem.

Fifty-four percent of eighth-graders selected the correct answer (choice B). The most common incorrect answer (choice A), which was selected by 17 percent of the students, resulted from interchanging the variables for the number of cards sold and the amount of profit. Incorrect choices C and D are alternate ways to represent the relationship between the number of cards sold and the profit on Monday, but they do not represent the relationship on the other days. Incorrect choice E can be obtained by interchanging the variables and considering only Thursday.

	Mon.	Tues.	Wed.	Thurs.	Fri.	Sat.
Number Sold, n	4	0	5	2	3	6
Profit, p	$2.00	$0.00	$2.50	$1.00	$1.50	$3.00

Angela makes and sells special-occasion greeting cards. The table above shows the relationship between the number of cards sold and her profi t. Based on the data in the table, which of the following equations shows how the number of cards sold and profi t (in dollars) are related?

oA p = 2n oB p = 0.5n oC p = n – 2 oD p = 6 – n oE p = n + 1

Percentage of eighth-grade students in each response category in 2007

Choice A	Choice B	Choice C	Choice D	Choice E	Omitted
17	54	13	9	6	1

Note: Detail may not sum to totals because of rounding.

The table below shows the percentage of eighth graders within each achievement level who answered this question correctly. For example, 46 percent of eighth-graders at the *Basic* level selected the correct answer choice.

Percentage correct for eighth-grade students at each achievement level in 2007

Overall	Below Basic	At Basic	At Profi cient	At Advanced
54	22	46	86	98

Source: U.S. Department of Education, Institute of Education Sciences, National Center for Education Statistics, National Assessment of Educational Progress (NAEP), 2007 Mathematics Assessment.

Sample Question About Number Properties and Operations

This sample question measures eighth-graders' understanding in the number properties and operations content area. It addresses the "Properties of number and operations" subtopic, which focuses on recognizing, describing, and explaining properties of integers and operations. The framework objective measured by this question is "Explain or justify a mathematical concept or relationship." Students were permitted to use a calculator to solve this problem.

Student responses for this question were rated using a two- level scoring guide, rating responses as "Correct" or "Incorrect."

Forty-two percent of grade 8 students correctly responded to this question. The student response on the right was rated as "Correct." It showed that if two of the three numbers are 23 and 62, then the third number must be 88. Therefore, 62 cannot be the largest of the three numbers.

Percentage of eighth-grade students in each response category in 2007

Correct	Incorrect	Omitted
42	55	2

Note: Detail may not sum to totals because a small percentage of responses that did not address the assessment task are not shown.

The table below shows the percentage of eighth-graders within each achievement level whose answer to this question was rated as "Correct." For example, 43 percent of eighth-graders at the *Basic* level provided a response rated as "Correct."

Percentage rated as "correct" for eighth-grade students at each achievement level in 2007

Overall	Below Basic	At Basic	At Profi cient	At Advanced
42	13	43	66	78

Source: U.S. Department of Education, Institute of Education Sciences, National Center for Education Statistics, National Assessment of Educational Progress (NAEP), 2007 Mathematics Assessment.

TECHNICAL NOTES

Sampling and Weighting

The schools and students participating in NAEP assessments are selected to be representative both nationally and for public schools at the state level. Samples of schools and students are drawn from each state and from the District of Columbia and Department of Defense schools. The results from the assessed students are combined to provide accurate estimates of the overall performance of students in the nation and in individual states and other jurisdictions.

While national results reflect the performance of students in both public schools and nonpublic schools (i.e., private schools, Bureau of Indian Education schools, and Department of Defense schools), state-level results reflect the performance of public school students only. More information on sampling can be found at http://nces.ed.gov/nationsreportcard/about/nathow.asp.

Each school that participated in the assessment, and each student assessed, represents a portion of the population of interest. Results are weighted to make appropriate inferences between the student samples and the respective populations from which they are drawn. Sampling weights account for the disproportionate representation of the selected sample. This includes oversampling of schools with high concentrations of students from certain minority groups and the lower sampling rates of students who attend very small nonpublic schools.

Interpreting Statistical Significance

Comparisons over time or between groups are based on statistical tests that consider both the size of the differences and the standard errors of the two statistics being compared. Standard errors are margins of error, and estimates based on smaller groups are likely to have larger margins of error. The size of the standard errors may also be in(uenced by other factors such as how representative the students assessed are of the entire population.

When an estimate has a large standard error, a numerical difference that seems large may not be statistically signi2 cant. Differences of the same magnitude may or may not be statistically signi2 cant depending upon the size of the standard errors of the estimates. For example, a 1-point difference between male and female students may be statistically significant, while a 1-point difference between black and Asian/Paci2 c Islander students may not be. Standard errors for the estimates presented in this report are available at http://nces.ed.gov/nationsreportcard/nde.

School and Student Participation Rates

To ensure unbiased samples, NCES and the Governing Board established participation rate standards that states and jurisdictions were required to meet in order for their results to be reported. Participation rates for the original sample needed to be at least 85 percent for schools to meet reporting requirements. In the 2007 mathematics assessment, all 52 states and jurisdictions met participation rate standards at both grades 4 and 8.

The national school participation rates for public and private schools combined were 98 percent for grade 4 and 97 percent for grade 8. Student participation rates were 95 percent for grade 4 and 92 percent for grade 8.

Participation rates needed to be 70 percent or higher to report results separately for private schools. While the school participation rate for private schools did meet the standard in 2007, it did not always meet the standard in previous assessment years. Therefore, comparisons could not be made for private schools as a group across all years. Participation rates for Catholic schools, however, were sufficient for reporting in 2007 and in previous assessment years. These data and other private school data are available at http:// nationsreportcard.gov/math_2007/m0038.asp.

National School Lunch Program

NAEP first began collecting data in 1996 on student eligibility for the National School Lunch Program (NSLP) as an indicator of poverty. Under the guidelines of NSLP, children from families with incomes below 130 percent of the poverty level are eligible for free meals. Those from families with incomes between 130 and 185 percent of the poverty level are eligible for reduced-price meals. (For the period July 1, 2006 through June 30, 2007, for a family of four, 130 percent of the poverty level was $26,000, and 185 percent was $37,000.)

As a result of improvements in the quality of the data on students' eligibility for NSLP, the percentage of students for whom information was not available has decreased in comparison to the percentages reported prior to the 2003 assessment. Therefore, trend comparisons are only made back to 2003 in this report. For more information on NSLP, visit http://www.fns.usda.gov/cnd/lunch/.

APPENDIX TABLES

Table A-1. Fourth- and eighth-grade public and nonpublic school students with disabilities (SD) and/or English language learners (ELL) identified, excluded, and assessed in NAEP mathematics, as a percentage of all students: Various years, 1992–2007

	Accommodations not permitted		Accommodations permitted			
Student characteristics	1992	1996	1996	2000	2003	2005
Grade 4						
SD and/or ELL						
Identifi ed	9	14	15	18	21	21
Excluded	6	6	4	4	4	3
Assessed	3	8	11	14	17	18
Without accommodations	3	8	7	9	9	9
With accommodations	†	†	5	5	8	9
SD						
Identifi ed	7	11	10	12	13	13
Excluded	4	5	3	3	3	2
Assessed	3	6	7	9	10	10
Without accommodations	3	6	4	5	4	3
With accommodations	†	†	4	4	6	7
ELL						
Identifi ed	3	3	6	7	10	10
Excluded	2	1	1	1	1	1
Assessed	1	2	5	6	8	8
Without accommodations	1	2	3	4	6	6
With accommodations	†	†	2	1	2	2
Grade 8						
SD and/or ELL						
Identifi ed	9	11	12	13	17	17
Excluded	6	4	3	4	3	3
Assessed	4	6	8	10	14	14
Without accommodations	4	6	6	7	7	6
With accommodations	†	†	3	3	6	8
SD						
Identifi ed	7	9	9	10	13	12
Excluded	4	4	3	3	3	3
Assessed	3	5	6	7	10	10
Without accommodations	3	5	4	5	4	3
With accommodations	†	†	2	2	6	7
ELL						
Identifi ed	2	3	3	4	6	6
Excluded	2	1	1	1	1	1
Assessed	1	2	2	3	5	5
Without accommodations	1	2	2	2	4	4
With accommodations	†	†	#	1	1	1

† Not applicable. Accommodations were not permitted in this sample.

Rounds to zero.

Note: Students identified as both SD and ELL were counted only once under the combined SD and/or ELL category, but were counted separately under the SD and ELL categories. Detail may not sum to totals because of rounding.

Source: U.S. Department of Education, Institute of Education Sciences, National Center for Education Statistics, National Assessment of Educational Progress (NAEP), various years, 1992–2007 Mathematics Assessments.

Table A-2. Fourth- and eighth-grade public and nonpublic school students with disabilities (SD) and/or English language learners (ELL) identified, excluded, and assessed in NAEP mathematics, as a percentage of all students, by selected race/ethnicity categories: 2007

	Race/ethnicity		
Student characteristics	White	Black	Hispanic
Grade 4			
SD and/or ELL			
Identifi ed	14	16	46
Excluded	2	4	4
Assessed	12	12	42
Without accommodations	4	3	26
With accommodations	8	9	15
SD			
Identifi ed	13	14	12
Excluded	2	4	3
Assessed	11	11	9
Without accommodations	4	2	3
With accommodations	8	8	6
ELL			
Identifi ed	1	2	39
Excluded	#	#	3
Assessed	1	2	37
Without accommodations	#	1	25
With accommodations	#	1	12
Grade 8			
SD and/or ELL			
Identifi ed	12	16	33
Excluded	3	6	5
Assessed	9	11	28
Without accommodations	3	3	18
With accommodations	6	8	11
SD			
Identifi ed	11	15	11
Excluded	3	6	3
Assessed	8	10	8
Without accommodations	2	2	3
With accommodations	6	8	5
ELL			
Identifi ed	1	1	26
Excluded	#	#	3
Assessed	1	1	23
Without accommodations	#	#	16
With accommodations	#	#	7

Rounds to zero.

Note: Black includes African American, and Hispanic includes Latino. Race categories exclude Hispanic origin. Students identified as both SD and ELL were counted only once under the combined SD and/or ELL category, but were counted separately under the SD and ELL categories. Detail may not sum to totals because of rounding. SOURCE: U.S. Department of Education, Institute of Education Sciences, National Center for Education Statistics, National Assessment of Educational Progress (NAEP), 2007 Mathematics Assessment

Table A-3. Fourth- and eighth-grade public school students with disabilities (SD) and English language learners (ELL) identified, excluded, and accommodated in NAEP mathematics, as a percentage of all students, by state: 2007

State/jurisdiction	Grade 4							Grade 8						
	Overall	SD			ELL			Overall	SD			ELL		
	excluded	Identified	Excluded	Accommodated	Identified	Excluded	Accommodated	exclude	Identified	Excluded	Accommodated	Identified	Excluded	Accommodated
Nation (public)	**3**	**14**	**3**	**8**	**11**	**1**	**3**	**4**	**13**	**4**	**6**	**7**	**1**	**2**
Alabama	2	11	1	4	2	#	#	3	12	3	2	2	#	#
Alaska	2	16	1	10	16	1	6	4	12	4	6	17	1	5
Arizona	3	11	2	5	16	2	3	3	11	3	5	10	1	2
Arkansas	3	12	2	7	7	1	5	2	12	2	8	3	#	2
California	2	10	2	4	34	1	3	2	9	2	3	22	1	2
Colorado	2	12	2	9	15	#	7	2	10	2	7	7	#	3
Connecticut	1	13	1	9	7	#	5	2	13	1	9	4	#	2
Delaware	5	17	5	9	5	1	2	7	14	6	6	3	1	1
Florida	3	15	2	12	8	2	5	3	13	2	10	6	1	4
Georgia	2	12	2	7	3	#	2	5	9	5	3	2	#	1
Hawaii	1	11	1	8	10	1	4	2	13	1	7	7	1	3
Idaho	2	11	1	6	8	#	2	2	10	1	5	6	#	2
Illinois	5	15	3	8	9	1	3	6	14	5	8	4	1	1
Indiana	3	17	3	9	5	#	3	6	15	5	8	4	#	1
Iowa	1	13	1	10	5	#	3	2	15	2	11	3	#	2
Kansas	3	13	3	7	8	#	4	4	12	4	7	4	#	1
Kentucky	3	15	2	7	2	#	1	7	13	6	5	2	#	1
Louisiana	2	18	2	13	1	#	1	3	12	3	8	1	#	1
Maine	3	18	3	11	2	#	1	5	17	5	9	2	#	#
Maryland	4	12	4	6	4	1	3	7	11	7	3	2	#	1
Massachusetts	5	18	5	11	6	1	2	9	17	9	6	3	1	1
Michigan	3	13	3	7	2	#	1	5	14	4	8	2	#	#
Minnesota	2	13	2	7	8	1	3	2	12	2	7	5	#	1
Mississippi	1	10	1	6	1	#	#	2	11	2	6	#	#	#
Missouri	4	15	3	7	2	#	1	5	13	5	6	2	#	1
Montana	2	13	2	8	4	#	2	3	13	3	8	5	#	2
Nebraska	3	17	2	9	8	1	2	3	13	2	7	3	1	1
Nevada	3	13	2	6	22	2	9	4	12	3	5	11	1	4
New Hampshire	2	19	2	13	3	#	1	3	19	3	12	2	#	1
New Jersey	2	14	2	11	4	#	3	3	14	3	11	4	1	2
New Mexico	4	13	3	7	23	2	9	3	12	2	7	17	2	4
New York	2	15	1	12	9	1	7	3	14	3	11	5	1	4
North Carolina	2	15	2	10	7	1	4	2	13	2	10	4	#	2
North Dakota	4	15	4	8	3	1	1	6	14	6	6	3	#	1
Ohio	5	15	4	8	3	1	1	7	15	7	7	1	#	#
Oklahoma	5	14	5	6	5	#	1	8	14	8	4	4	1	1
Oregon	3	15	2	8	13	1	7	3	12	3	5	9	1	3
Pennsylvania	2	17	2	10	2	#	1	4	15	4	9	2	1	1
Rhode Island	2	19	2	12	7	1	4	3	17	2	12	4	1	1
South Carolina	2	13	2	6	4	#	1	5	13	5	5	2	#	1
South Dakota	1	15	1	7	4	#	1	2	11	2	6	1	#	#
Tennessee	6	14	6	4	2	#	1	6	12	6	3	2	#	1
Texas	5	13	5	5	16	2	5	6	11	5	3	8	2	2
Utah	2	12	2	6	12	1	4	3	10	2	6	9	1	2
Vermont	2	17	2	11	3	#	1	4	19	4	10	2	#	1
Virginia	5	15	4	7	8	1	4	7	14	6	6	4	1	1
Washington	3	15	2	8	9	1	4	4	11	3	6	6	1	2
West Virginia	1	17	1	8	1	#	#	2	17	2	10	1	#	#
Wisconsin	3	15	2	9	7	1	4	5	14	4	9	5	1	2
Wyoming	2	15	2	9	4	#	1	2	13	2	9	3	#	1
Other jurisdictions														
District of Columbia	6	14	5	8	8	2	5	10	17	9	6	4	1	2
DoDEA[1]	2	11	1	7	7	1	2	2	7	1	6	5	1	1

Rounds to zero.

[1] Department of Defense Education Activity (overseas and domestic schools).

Note: Students identified as both SD and ELL were counted only once in overall, but were counted separately under the SD and ELL categories.

Source: U.S. Department of Education, Institute of Education Sciences, National Center for Education Statistics, National Assessment of Educational Progress (NAEP), 2007 Mathematics Assessment

Table A-4. Fourth- and eighth-grade public school students with disabilities excluded in NAEP mathematics, as a percentage of all students, by state: Various years, 1990–2007

	Grade 4						Grade 8						
State/jurisdiction	1992[1]	1996[1]	2000	2003	2005	2007	1990[1]	1992[1]	1996[1]	2000	2003	2005	2007
Nation (public)	**5**	**5**	**3**	**3**	**3**	**3**	**—**	**5**	**4**	**3**	**3**	**3**	**4**
Alabama	4	6	3	2	1	1	5	5	7	6	2	1	3
Alaska	—	4	—	1	1	1	—	—	5	—	1	2	4
Arizona	3	7	3	3	3	2	3	4	5	2	3	3	3
Arkansas	5	6	4	1	2	2	7	6	7	2	1	3	2
California	3	5	3	2	2	2	3	4	5	3	1	2	2
Colorado	4	7	—	2	2	2	4	4	4	—	1	2	2
Connecticut	4	7	3	3	2	1	5	5	7	5	3	2	1
Delaware	5	6	—	6	7	5	4	4	8	—	8	10	6
Florida	7	7	—	2	2	2	5	5	7	—	2	2	2
Georgia	5	6	3	2	2	2	3	4	6	4	2	2	5
Hawaii	5	4	6	2	2	1	3	3	4	4	3	2	1
Idaho	3	—	1	1	1	1	2	3	—	2	1	2	1
Illinois	—	—	2	3	2	3	4	—	—	3	4	3	5
Indiana	3	5	2	2	1	3	5	4	5	3	2	4	5
Iowa	3	5	1	2	2	1	4	4	5	—	2	2	2
Kansas	—	—	3	1	2	3	—	—	—	3	2	3	4
Kentucky	3	6	3	3	2	2	5	5	4	4	4	3	6
Louisiana	4	7	3	3	4	2	4	4	6	2	4	4	3
Maine	6	7	4	3	3	3	—	4	5	3	4	4	5
Maryland	3	7	2	3	3	4	4	4	6	2	3	4	7
Massachusetts	6	7	1	2	3	5	—	6	7	2	2	6	9
Michigan	5	6	3	3	4	3	4	6	5	4	4	4	4
Minnesota	3	5	2	2	2	2	3	3	3	1	2	2	2
Mississippi	5	6	3	5	2	1	—	7	7	5	5	3	2
Missouri	4	5	2	3	2	3	—	4	6	3	4	4	5
Montana	—	5	2	2	2	2	2	—	3	2	2	2	3
Nebraska	4	4	2	2	2	2	3	4	4	3	3	1	2
Nevada	—	5	3	3	3	2	—	—	5	3	2	2	3
New Hampshire	4	—	—	3	2	2	4	5	4	—	3	2	3
New Jersey	3	5	—	2	2	2	5	6	5	—	1	3	3
New Mexico	6	8	5	2	2	3	6	4	5	7	2	2	2
New York	3	5	2	3	3	1	4	6	5	3	4	3	3
North Carolina	3	6	4	4	2	2	3	3	4	4	3	2	2
North Dakota	2	3	1	2	2	4	2	2	3	2	1	4	6
Ohio	6	—	4	4	3	4	5	6	—	4	5	5	7
Oklahoma	7	—	4	3	4	5	5	6	—	4	2	4	8
Oregon	—	6	2	4	3	2	2	—	3	2	3	2	3
Pennsylvania	3	4	—	2	2	2	5	4	—	—	1	3	4
Rhode Island	4	5	2	2	2	2	5	4	5	3	3	3	2
South Carolina	5	5	5	6	4	2	—	6	6	4	7	6	5
South Dakota	—	—	—	1	1	1	—	—	—	—	2	2	2
Tennessee	4	6	2	2	3	6	—	5	4	2	3	5	6
Texas	5	7	6	7	5	5	4	5	6	7	6	5	5
Utah	4	5	3	2	2	2	—	4	5	2	2	2	2
Vermont	—	6	3	4	3	2	—	—	4	3	3	4	4
Virginia	5	6	3	4	4	4	4	5	7	5	6	4	6
Washington	—	5	—	2	2	2	—	—	5	—	2	2	3
West Virginia	4	8	3	3	2	1	5	6	8	3	3	3	2
Wisconsin	5	7	4	3	2	2	4	4	7	4	3	3	4
Wyoming	3	4	2	1	1	2	3	4	2	1	1	2	2
Other jurisdictions													
District of Columbia	7	7	3	4	5	5	4	8	8	5	5	5	9
DoDEA[2]	—	4	2	1	1	1	—	—	2	1	1	1	1

— Not available. The jurisdiction did not participate or did not meet the minimum participation guidelines for reporting.

[1] Accommodations were not permitted in this assessment year.

[2] Department of Defense Education Activity (overseas and domestic schools). Before 2005, DoDEA overseas and domestic schools were separate jurisdictions in NAEP. Pre-2005 data presented here were recalculated for comparability.

Source: U.S. Department of Education, Institute of Education Sciences, National Center for Education Statistics, National Assessment of Educational Progress (NAEP), various years, 1990–2007 Mathematics Assessments.

Table A-5. Fourth- and eighth-grade public school English language learners excluded in NAEP mathematics, as a percentage of all students, by state: Various years, 1990–2007

	Grade 4						Grade 8						
State/jurisdiction	1992[1]	1996[1]	2000	2003	2005	2007	1990[1]	1992[1]	1996[1]	2000	2003	2005	2007
Nation (public)	**2**	**2**	**1**	**1**	**1**	**1**	**—**	**2**	**1**	**1**	**1**	**1**	**1**
Alabama	#	#	#	#	#	#	#	#	#	#	#	#	#
Alaska	—	1	—	#	1	1	—	—	1	—	#	#	1
Arizona	2	7	3	2	2	2	1	2	4	1	2	2	1
Arkansas	#	#	#	1	2	1	#	#	#	#	1	1	#
California	10	12	3	2	3	1	4	5	6	2	2	1	1
Colorado	1	2	—	1	1	#	1	1	1	—	1	1	#
Connecticut	2	2	1	1	1	#	1	1	2	2	1	#	#
Delaware	1	1	—	1	1	1	#	#	#	—	1	1	1
Florida	2	3	—	2	1	2	2	2	3	—	1	1	1
Georgia	1	2	1	1	1	#	#	#	1	1	1	#	#
Hawaii	2	1	3	2	1	1	1	2	1	1	1	1	1
Idaho	1	—	2	1	1	#	#	#	—	1	#	1	#
Illinois	—	—	2	2	1	1	1	—	—	2	1	1	1
Indiana	#	#	1	#	1	#	#	#	#	#	#	#	#
Iowa	#	1	1	1	#	#	#	#	#	—	#	#	#
Kansas	—	—	#	#	1	#	—	—	—	#	1	1	#
Kentucky	#	#	#	1	#	#	#	#	#	1	1	#	#
Louisiana	#	1	#	#	#	#	#	#	#	#	1	#	#
Maine	#	#	#	1	#	#	—	#	#	#	#	#	#
Maryland	1	1	1	2	1	1	1	1	1	1	1	#	#
Massachusetts	1	2	2	1	1	1	—	2	1	2	1	1	1
Michigan	1	1	1	1	1	#	#	#	1	#	1	#	#
Minnesota	#	1	1	1	1	1	#	#	#	1	1	1	#
Mississippi	#	#	#	1	#	#	—	#	#	#	#	#	#
Missouri	#	#	1	1	#	#	—	#	1	#	#	#	#
Montana	—	#	#	#	#	#	#	—	#	#	#	#	#
Nebraska	#	1	1	1	1	1	#	#	1	1	1	#	1
Nevada	—	4	4	2	1	2	—	—	3	1	1	1	1
New Hampshire	#	—	—	1	#	#	#	#	#	—	#	#	#
New Jersey	2	1	—	1	1	#	2	1	2	—	1	1	1
New Mexico	1	5	2	2	1	2	1	1	4	2	1	2	2
New York	2	3	3	3	1	1	2	3	3	2	2	1	1
North Carolina	#	1	1	1	1	1	#	#	1	1	1	1	#
North Dakota	#	#	#	#	#	1	#	#	#	#	#	#	#
Ohio	#	—	#	1	#	1	#	#	—	1	#	#	#
Oklahoma	#	—	1	1	1	#	#	#	—	#	1	1	1
Oregon	—	3	1	1	1	1	#	—	1	1	1	1	1
Pennsylvania	1	1	—	1	#	#	#	#	—	—	#	#	1
Rhode Island	3	2	1	2	1	1	2	2	2	1	2	1	1
South Carolina	#	#	1	#	#	#	—	#	#	#	#	#	#
South Dakota	—	—	—	#	#	#	—	—	—	—	#	#	#
Tennessee	#	1	1	#	1	#	—	#	#	1	1	#	#
Texas	4	5	2	2	2	2	2	2	3	2	2	2	2
Utah	1	1	1	1	1	1	—	1	1	#	1	1	1
Vermont	—	#	#	#	#	#	—	—	#	1	#	#	#
Virginia	1	1	2	2	1	1	1	1	1	1	2	1	1
Washington	—	1	—	1	1	1	—	—	1	—	1	1	1
West Virginia	#	#	#	#	#	#	#	#	#	#	#	#	#
Wisconsin	1	1	1	1	1	1	#	#	1	1	1	1	1
Wyoming	#	#	#	#	#	#	#	#	#	#	#	#	#
Other jurisdictions													
District of Columbia	2	4	2	1	1	2	1	2	3	2	1	1	1
DoDEA[2]	—	1	1	1	1	1	—	—	1	1	1	1	1

— Not available. The jurisdiction did not participate or did not meet the minimum participation guidelines for reporting.

Rounds to zero.

[1] Accommodations were not permitted in this assessment year.

[2] Department of Defense Education Activity (overseas and domestic schools). Before 2005, DoDEA overseas and domestic schools were separate jurisdictions in NAEP. Pre-2005 data presented here were recalculated for comparability.

Source: U.S. Department of Education, Institute of Education Sciences, National Center for Education Statistics, National Assessment of Educational Progress (NAEP), various years, 1990–2007 Mathematics Assessments.

Table A-6. Percentage distribution of fourth- and eighth-grade students in NAEP mathematics, by selected race/ethnicity categories and state: 1990, 1992, and 2007

	Grade 4						Grade 8					
	White		Black		Hispanic		White		Black		Hispanic	
State/jurisdiction	1992	2007	1992	2007	1992	2007	1990	2007	1990	2007	1990	2007
Nation (public)[1]	**72***	**55**	**18***	**17**	**7***	**21**	**73***	**58**	**16**	**17**	**7***	**19**
Alabama	65	58	34	37	#*	3	68*	60	32	35	#*	2
Alaska	—	55	—	5	—	4	—	56	—	4	—	4
Arizona	62*	43	4	5	23*	44	62*	47	3	5	26*	39
Arkansas	75*	67	24	22	#*	9	75	69	24	22	1*	7
California	50*	27	7	7	30*	54	49*	31	7	7	30*	48
Colorado	73*	60	6	6	17*	30	77*	65	5	7	15*	25
Connecticut	77*	64	11	13	10*	18	79*	69	11	13	8*	15
Delaware	70*	54	25*	33	2*	10	70*	56	26*	31	2*	9
Florida	63*	48	24	21	12*	25	65*	48	22	23	12*	24
Georgia	60*	46	38	38	1*	9	62*	46	36*	43	1*	7
Hawaii	23*	17	3	3	2*	4	20*	14	2	2	2	2
Idaho	92*	81	#*	1	6*	13	93*	82	#	1	4*	14
Illinois	—	56	—	19	—	19	70*	60	19	16	8*	18
Indiana	87*	78	11	10	2*	7	87*	77	9	12	2*	7
Iowa	95*	86	2*	5	1*	6	95*	88	2*	4	1*	6
Kansas	—	73	—	8	—	13	—	76	—	8	—	10
Kentucky	90*	84	9	11	#*	2	90*	86	9	10	#*	2
Louisiana	53	47	45	49	1*	2	57	52	40	43	1	2
Maine	98*	95	#*	2	#*	1	—	96	—	2	—	1
Maryland	62*	50	32	35	2*	8	62*	51	31	37	2*	7
Massachusetts	83*	75	8	7	4*	11	—	75	—	8	—	10
Michigan	79*	71	16	21	3	3	82*	75	14	18	2*	3
Minnesota	91*	78	3*	8	2*	7	93*	81	2*	7	#*	4
Mississippi	42	45	58	52	#	2	—	47	—	51	—	1
Missouri	83*	77	15	19	1*	3	—	75	—	19	—	3
Montana	—	83	—	1	—	3	91*	85	#	1	1*	2
Nebraska	90*	75	6	7	3*	14	92*	80	5*	7	2*	11
Nevada	—	43	—	8	—	40	—	47	—	10	—	34
New Hampshire	96*	91	1*	2	1*	4	98*	94	#*	2	1*	3
New Jersey	69*	57	16	14	11*	20	69*	57	17	17	9*	19
New Mexico	45*	29	4	3	45*	58	42*	32	2	3	42*	52
New York	63*	53	15	19	17	20	61	55	19	19	13	18
North Carolina	65*	55	31	28	1*	10	63*	56	32	30	1*	8
North Dakota	95*	87	#*	2	1*	2	93	89	#	1	1	1
Ohio	86*	75	12*	18	1*	3	84*	76	12*	18	1*	2
Oklahoma	77*	58	9	11	3*	9	77*	59	11	9	2*	8
Oregon	—	71	—	3	—	17	91*	73	2*	3	3*	15
Pennsylvania	81	77	14	14	3	6	82	76	14	15	2*	6
Rhode Island	82*	70	7	8	7*	19	86*	70	5*	9	5*	17
South Carolina	58	57	41	36	#*	4	—	56	—	38	—	3
South Dakota	—	83	—	2	—	2	—	86	—	1	—	2
Tennessee	73	69	25	26	#*	3	—	67	—	28	—	4
Texas	49*	36	14	15	34*	45	50*	38	14	15	33*	44
Utah	93*	80	1	1	4*	15	—	82	—	1	—	12
Vermont	—	94	—	2	—	1	—	95	—	1	—	1
Virginia	71*	58	25	26	2*	8	70*	61	25	26	2*	6
Washington	—	65	—	6	—	15	—	69	—	5	—	14
West Virginia	96*	93	2*	5	#	1	96	94	3	4	#	1
Wisconsin	87*	77	6*	10	2*	8	88*	80	9	10	1*	6
Wyoming	90*	84	1	2	6*	10	86	86	1	1	6*	8
Other jurisdictions												
District of Columbia	5*	6	91*	84	3*	9	3	3	93*	88	3*	9
DoDEA[2]	—	51	—	17	—	14	—	48	—	18	—	15

— Not available. The jurisdiction did not participate or did not meet the minimum participation guidelines for reporting.

Rounds to zero.

* Significantly different (*p* < .05) from 2007 when only one jurisdiction or the nation is being examined.

[1] National results for assessments prior to 2003 are based on the national sample, not on aggregated state samples.

[2] Department of Defense Education Activity (overseas and domestic schools).

Note: Black includes African American, and Hispanic includes Latino. Race categories exclude Hispanic origin. State-level data were not collected at grade 4 in 1990.

Source: U.S. Department of Education, Institute of Education Sciences, National Center for Education Statistics, National Assessment of Educational Progress (NAEP), 1990, 1992, and 2007 Mathematics Assessments.

Table A-7. Percentage of fourth-grade public school students at or above *Basic* in NAEP mathematics, by state: Various years, 1992–2007

	Accommodations not permitted			Accommodations permitted			
State/jurisciction	1992	1996	2000	2000	2003	2005	2007
Nation (public)[1]	**57***	**62***	**67***	**64***	**76***	**79***	**81**
Alabama	43*	48*	57*	55*	65*	66	70
Alaska	—	65*	—	—	75*	77	79
Arizona	53*	57*	58*	57*	70	70	74
Arkansas	47*	54*	56*	55*	71*	78	81
California	46*	46*	52*	50*	67	71	70
Colorado	61*	67*	—	—	77*	81	82
Connecticut	67*	75*	77*	76*	82	84	84
Delaware	55*	54*	—	—	81*	84*	87
Florida	52*	55*	—	—	76*	82*	86
Georgia	53*	53*	58*	57*	72*	76	79
Hawaii	52*	53*	55*	55*	68*	73*	77
Idaho	63*	—	71*	68*	80*	86	85
Illinois	—	—	66*	63*	73*	74*	79
Indiana	60*	72*	78*	77*	82*	84*	89
Iowa	72*	74*	78*	75*	83*	85	87
Kansas	—	—	75*	76*	85*	88	89
Kentucky	51*	60*	60*	59*	72*	75*	79
Louisiana	39*	44*	57*	57*	67*	74	73
Maine	75*	75*	74*	73*	83	84	85
Maryland	55*	59*	61*	60*	73*	79	80
Massachusetts	68*	71*	79*	77*	84*	91*	93
Michigan	61*	68*	72*	71*	77	79	80
Minnesota	71*	76*	78*	76*	84*	88	87
Mississippi	36*	42*	45*	45*	62*	69	70
Missouri	62*	66*	72*	71*	79	79*	82
Montana	—	71*	73*	72*	81*	85	88
Nebraska	67*	70*	67*	65*	80	80	80
Nevada	—	57*	61*	60*	69*	72	74
New Hampshire	72*	—	—	—	87*	89	91
New Jersey	68*	68*	—	—	80*	86*	90
New Mexico	50*	51*	51*	50*	63*	65*	70
New York	57*	64*	67*	66*	79*	81*	85
North Carolina	50*	64*	76*	73*	85	83	85
North Dakota	72*	75*	75*	73*	83*	89	91
Ohio	57*	—	73*	73*	81*	84*	87
Oklahoma	60*	—	69*	67*	74*	79*	82
Oregon	—	65*	67*	65*	79	80	79
Pennsylvania	65*	68*	—	—	78*	82	85
Rhode Island	54*	61*	67*	65*	72*	76	80
South Carolina	48*	48*	60*	59*	79	81	80
South Dakota	—	—	—	—	82*	86	86
Tennessee	47*	58*	60*	59*	70*	74	76
Texas	56*	69*	77*	76*	82*	87	87
Utah	66*	69*	70*	69*	79*	83	83
Vermont	—	67*	73*	73*	85*	87*	89
Virginia	59*	62*	73*	71*	83*	83*	87
Washington	—	67*	—	—	81*	84	84
West Virginia	52*	63*	68*	65*	75*	75*	81
Wisconsin	71*	74*	—	—	79*	84	85
Wyoming	69*	64*	73*	71*	87	87	88
Other jurisdictions							
District of Columbia	23*	20*	24*	24*	36*	45*	49
DoDEA[2]	—	64*	70*	69*	84	85	86

— Not available. The jurisdiction did not participate or did not meet the minimum participation guidelines for reporting.

* Significantly different ($p < .05$) from 2007 when only one jurisdiction or the nation is being examined.

[1] National results for assessments prior to 2003 are based on the national sample, not on aggregated state samples.

[2] Department of Defense Education Activity (overseas and domestic schools). Before 2005, DoDEA overseas and domestic schools were separate jurisdictions in NAEP. Pre-2005 data presented here were recalculated for comparability.

Note: State-level data were not collected in 1990.

Source: U.S. Department of Education, Institute of Education Sciences, National Center for Education Statistics, National Assessment of Educational Progress (NAEP), various years, 1992–2007 Mathematics Assessments.

Table A-8. Percentage of fourth-grade public school students at or above *Proficient* in NAEP mathematics, by state: Various years, 1992–2007

State/jurisdiction	Accommodations not permitted			Accommodations permitted			
	1992	1996	2000	2000	2003	2005	2007
Nation (public)[1]	**17***	**20***	**25***	**22***	**31***	**35***	**39**
Alabama	10*	11*	14*	13*	19*	21*	26
Alaska	—	21*	—	—	30*	34	38
Arizona	13*	15*	17*	16*	25*	28	31
Arkansas	10*	13*	13*	14*	26*	34	37
California	12*	11*	15*	13*	25*	28	30
Colorado	17*	22*	—	—	34*	39	41
Connecticut	24*	31*	32*	31*	41	42	45
Delaware	17*	16*	—	—	31*	36*	40
Florida	13*	15*	—	—	31*	37*	40
Georgia	15*	13*	18*	17*	27*	30	32
Hawaii	15*	16*	14*	14*	23*	27*	33
Idaho	16*	—	21*	20*	31*	40	40
Illinois	—	—	21*	20*	32*	32*	36
Indiana	16*	24*	31*	30*	35*	38*	46
Iowa	26*	22*	28*	26*	36*	37*	43
Kansas	—	—	30*	29*	41*	47	51
Kentucky	13*	16*	17*	17*	22*	26*	31
Louisiana	8*	8*	14*	14*	21	24	24
Maine	27*	27*	25*	23*	34*	39	42
Maryland	18*	22*	22*	21*	31*	38	40
Massachusetts	23*	24*	33*	31*	41*	49*	58
Michigan	18*	23*	29*	28*	34	38	37
Minnesota	26*	29*	34*	33*	42*	47	51
Mississippi	6*	8*	9*	9*	17*	19	21
Missouri	19*	20*	23*	23*	30*	31*	38
Montana	—	22*	25*	24*	31*	38*	44
Nebraska	22*	24*	24*	24*	34*	36	38
Nevada	—	14*	16*	16*	23*	26*	30
New Hampshire	25*	—	—	—	43*	47*	52
New Jersey	25*	25*	—	—	39*	45*	52
New Mexico	11*	13*	12*	12*	17*	19*	24
New York	17*	20*	22*	21*	33*	36*	43
North Carolina	13*	21*	28*	25*	41	40	41
North Dakota	22*	24*	25*	25*	34*	40*	46
Ohio	16*	—	26*	25*	36*	43	46
Oklahoma	14*	—	16*	16*	23*	29	33
Oregon	—	21*	23*	23*	33	37	35
Pennsylvania	22*	20*	—	—	36*	41*	47
Rhode Island	13*	17*	23*	22*	28*	31*	34
South Carolina	13*	12*	18*	18*	32*	36	36
South Dakota	—	—	—	—	34*	41	41
Tennessee	10*	17*	18*	18*	24*	28	29
Texas	15*	25*	27*	25*	33*	40	40
Utah	19*	23*	24*	23*	31*	37	39
Vermont	—	23*	29*	29*	42*	44*	49
Virginia	19*	19*	25*	24*	36*	39	42
Washington	—	21*	—	—	36*	42	44
West Virginia	12*	19*	18*	17*	24*	25*	33
Wisconsin	24*	27*	—	—	35*	40*	47
Wyoming	19*	19*	25*	25*	39*	43	44
Other jurisdictions							
District of Columbia	5*	5*	6*	5*	7*	10*	14
DoDEA[2]	—	19*	23*	21*	31*	35	37

— Not available. The jurisdiction did not participate or did not meet the minimum participation guidelines for reporting.

* Significantly different ($p < .05$) from 2007 when only one jurisdiction or the nation is being examined.

[1] National results for assessments prior to 2003 are based on the national sample, not on aggregated state samples.

[2] Department of Defense Education Activity (overseas and domestic schools). Before 2005, DoDEA overseas and domestic schools were separate jurisdictions in NAEP. Pre-2005 data presented here were recalculated for comparability.

Note: State-level data were not collected in 1990.

Source: U.S. Department of Education, Institute of Education Sciences, National Center for Education Statistics, National Assessment of Educational Progress (NAEP), various years, 1992–2007 Mathematics Assessments.

Table A-9. Average scale scores and achievement-level results in NAEP mathematics for fourth-grade public school students, by race/ethnicity and state: 2007

	White					Black					Hispanic				
		Percentage of students					Percentage of students					Percentage of students			
State/jurisdiction	Average scale	Below Basic	At or above Basic	At or above Proficient	At Advanced	Average scale	Below Basic	At or above Basic	At or above Proficient	At Advanced	Average scale	Below Basic	At or above Basic	At or above Proficient	At Advanced
Nation (public)	**248**	**9**	**91**	**51**	**8**	**222**	**37**	**63**	**15**	**1**	**227**	**31**	**69**	**22**	**1**
Alabama	238	17	83	36	4	213	50	50	10	1	218	45	55	17	1
Alaska	247	10	90	50	8	227	33	67	22	2	232	24	76	26	2
Arizona	246	11	89	48	8	219	41	59	16	1	220	39	61	15	#
Arkansas	245	11	89	46	6	217	44	56	12	#	230	23	77	22	1
California	247	12	88	52	9	218	42	58	15	1	218	43	57	15	1
Colorado	249	9	91	54	9	224	35	65	20	2	224	34	66	19	2
Connecticut	252	6	94	57	10	220	40	60	15	1	223	36	64	18	2
Delaware	249	6	94	53	7	230	24	76	20	#	234	17	83	25	1
Florida	250	6	94	54	8	225	29	71	15	1	238	17	83	33	3
Georgia	246	10	90	46	6	222	36	64	13	1	229	25	75	20	1
Hawaii	244	14	86	46	7	230	25	75	24	3	224	33	67	19	2
Idaho	245	11	89	45	6	‡	‡	‡	‡	‡	224	36	64	18	1
Illinois	248	9	91	50	8	216	46	54	9	#	223	36	64	19	1
Indiana	249	8	92	52	7	224	30	70	14	1	233	20	80	26	1
Iowa	245	11	89	46	6	224	34	66	17	1	230	29	71	25	3
Kansas	252	7	93	58	10	226	29	71	21	#	234	22	78	29	2
Kentucky	238	18	82	34	4	219	41	59	12	#	221	38	62	15	1
Louisiana	240	14	86	37	4	219	40	60	11	#	234	23	77	31	3
Maine	243	14	86	43	6	221	38	62	17	#	‡	‡	‡	‡	‡
Maryland	251	9	91	55	12	223	37	63	17	1	233	24	76	28	3
Massachusetts	257	3	97	65	12	232	25	75	26	2	231	23	77	23	2
Michigan	244	12	88	44	6	216	48	52	12	#	230	28	72	26	2
Minnesota	252	8	92	58	11	222	38	62	16	1	229	28	72	22	2
Mississippi	239	13	87	34	2	217	45	55	9	#	‡	‡	‡	‡	‡
Missouri	245	12	88	45	6	218	43	57	12	1	234	22	78	26	3
Montana	247	9	91	49	6	‡	‡	‡	‡	‡	241	15	85	40	4
Nebraska	244	12	88	45	6	211	56	44	9	1	220	40	60	15	1
Nevada	243	13	87	43	5	219	42	58	16	1	221	39	61	18	1
New Hampshire	250	7	93	53	8	226	33	67	25	#	232	25	75	27	#
New Jersey	255	5	95	63	11	232	22	78	25	2	234	21	79	29	3
New Mexico	242	14	86	43	5	220	39	61	18	#	222	37	63	16	1
New York	251	6	94	56	8	225	31	69	18	1	230	26	74	25	2
North Carolina	251	6	94	56	9	224	32	68	15	1	235	16	84	28	2
North Dakota	248	6	94	49	5	‡	‡	‡	‡	‡	‡	‡	‡	‡	‡
Ohio	250	7	93	53	8	225	33	67	18	1	231	24	76	25	1
Oklahoma	242	12	88	39	4	220	37	63	10	#	227	30	70	22	1
Oregon	241	15	85	40	5	219	41	59	16	1	217	46	54	12	1
Pennsylvania	249	10	90	53	8	222	36	64	18	1	229	30	70	28	3
Rhode Island	242	14	86	41	4	219	41	59	16	1	220	38	62	15	#
South Carolina	248	10	90	50	8	221	36	64	14	1	227	26	74	21	2
South Dakota	245	9	91	46	4	221	37	63	15	2	228	31	69	21	2
Tennessee	240	14	86	36	4	214	50	50	9	#	222	33	67	15	1
Texas	253	5	95	59	9	230	24	76	21	1	236	16	84	30	2
Utah	244	12	88	45	5	‡	‡	‡	‡	‡	220	42	58	16	1
Vermont	247	10	90	50	8	‡	‡	‡	‡	‡	‡	‡	‡	‡	‡
Virginia	251	7	93	53	9	228	27	73	18	1	235	18	82	28	1
Washington	248	10	90	51	8	222	37	63	17	2	225	32	68	19	1
West Virginia	237	18	82	33	3	223	36	64	19	1	‡	‡	‡	‡	‡
Wisconsin	250	8	92	54	8	212	53	47	10	1	229	31	69	27	1
Wyoming	246	9	91	48	5	‡	‡	‡	‡	‡	229	27	73	23	1
Other jurisdictions															
District of Columbia	262	9	91	73	27	209	55	45	8	#	220	43	57	19	1
DoDEA[1]	246	8	92	47	5	227	28	72	17	#	233	20	80	25	1

See notes at end of table.

Table A-9. Continued

State/jurisdiction	Asian/Pacific Islander: Average scale score	Asian/Pacific Islander: Percentage of students Below *Basic*	At or above *Basic*	At or above *Proficient*	At *Advanced*	American Indian/Alaska Native: Average scale score	American Indian/Alaska Native: Percentage of students Below *Basic*	At or above *Basic*	At or above *Proficient*	At *Advanced*
Nation (public)	**254**	**9**	**91**	**59**	**16**	**229**	**28**	**72**	**26**	**3**
Alabama	‡	‡	‡	‡	‡	‡	‡	‡	‡	‡
Alaska	237	21	79	37	4	218	43	57	16	2
Arizona	253	9	91	59	15	216	45	55	15	1
Arkansas	236	23	77	41	7	‡	‡	‡	‡	‡
California	251	11	89	56	15	‡	‡	‡	‡	‡
Colorado	247	12	88	53	9	‡	‡	‡	‡	‡
Connecticut	255	8	92	64	17	‡	‡	‡	‡	‡
Delaware	261	1	99	70	17	‡	‡	‡	‡	‡
Florida	255	7	93	59	17	‡	‡	‡	‡	‡
Georgia	255	10	90	63	14	‡	‡	‡	‡	‡
Hawaii	233	24	76	31	4	‡	‡	‡	‡	‡
Idaho	‡	‡	‡	‡	‡	215	45	55	13	2
Illinois	257	5	95	62	17	‡	‡	‡	‡	‡
Indiana	‡	‡	‡	‡	‡	‡	‡	‡	‡	‡
Iowa	‡	‡	‡	‡	‡	‡	‡	‡	‡	‡
Kansas	260	7	93	67	21	‡	‡	‡	‡	‡
Kentucky	‡	‡	‡	‡	‡	‡	‡	‡	‡	‡
Louisiana	‡	‡	‡	‡	‡	‡	‡	‡	‡	‡
Maine	‡	‡	‡	‡	‡	‡	‡	‡	‡	‡
Maryland	261	7	93	68	23	‡	‡	‡	‡	‡
Massachusetts	259	5	95	66	21	‡	‡	‡	‡	‡
Michigan	261	4	96	69	23	‡	‡	‡	‡	‡
Minnesota	239	21	79	43	6	234	22	78	28	5
Mississippi	‡	‡	‡	‡	‡	‡	‡	‡	‡	‡
Missouri	‡	‡	‡	‡	‡	‡	‡	‡	‡	‡
Montana	‡	‡	‡	‡	‡	222	36	64	16	1
Nebraska	‡	‡	‡	‡	‡	‡	‡	‡	‡	‡
Nevada	242	15	85	43	4	‡	‡	‡	‡	‡
New Hampshire	258	8	92	64	20	‡	‡	‡	‡	‡
New Jersey	267	2	98	78	26	‡	‡	‡	‡	‡
New Mexico	‡	‡	‡	‡	‡	222	38	62	17	1
New York	260	6	94	69	21	‡	‡	‡	‡	‡
North Carolina	253	9	91	60	14	229	27	73	24	3
North Dakota	‡	‡	‡	‡	‡	224	34	66	17	#
Ohio	‡	‡	‡	‡	‡	‡	‡	‡	‡	‡
Oklahoma	247	8	92	48	6	234	20	80	29	2
Oregon	249	12	88	53	14	220	39	61	18	2
Pennsylvania	259	5	95	66	18	‡	‡	‡	‡	‡
Rhode Island	244	12	88	41	8	‡	‡	‡	‡	‡
South Carolina	‡	‡	‡	‡	‡	‡	‡	‡	‡	‡
South Dakota	‡	‡	‡	‡	‡	218	40	60	13	#
Tennessee	‡	‡	‡	‡	‡	‡	‡	‡	‡	‡
Texas	263	1	99	70	23	‡	‡	‡	‡	‡
Utah	244	11	89	44	5	‡	‡	‡	‡	‡
Vermont	‡	‡	‡	‡	‡	‡	‡	‡	‡	‡
Virginia	256	4	96	60	15	‡	‡	‡	‡	‡
Washington	250	12	88	54	14	227	32	68	26	4
West Virginia	‡	‡	‡	‡	‡	‡	‡	‡	‡	‡
Wisconsin	245	16	84	50	8	‡	‡	‡	‡	‡
Wyoming	‡	‡	‡	‡	‡	227	26	74	21	#
Other jurisdictions										
District of Columbia	‡	‡	‡	‡	‡	‡	‡	‡	‡	‡
DoDEA[1]	239	15	85	36	2	‡	‡	‡	‡	‡

Rounds to zero.

‡ Reporting standards not met. Sample size is insufficient to permit a reliable estimate.

[1] Department of Defense Education Activity (overseas and domestic schools).

Note: Black includes African American, Hispanic includes Latino, and Pacific Islander includes Native Hawaiian. Race categories exclude Hispanic origin. Results are not shown for students whose race/ethnicity was "unclassified." Detail may not sum to totals because of rounding.

Source: U.S. Department of Education, Institute of Education Sciences, National Center for Education Statistics, National Assessment of Educational Progress (NAEP), 2007 Mathematics Assessment.

Table A-10. Average scale scores and achievement-level results in NAEP mathematics for fourth-grade public school students, by gender and state: 2007

	Male					Female				
		Percentage of students					Percentage of students			
State/jurisdiction	Average scale score	Below *Basic*	At or above *Basic*	At or above *Profi cient*	At *Advanced*	Average scale score	Below *Basic*	At or above *Basic*	At or above *Profi cient*	At *Advanced*
Nation (public)	**240**	**18**	**82**	**41**	**7**	**238**	**19**	**81**	**36**	**4**
Alabama	229	30	70	27	3	228	30	70	25	2
Alaska	238	21	79	38	7	237	21	79	37	5
Arizona	233	26	74	34	5	230	27	73	27	3
Arkansas	238	20	80	38	5	237	19	81	35	4
California	231	30	70	31	5	229	31	69	28	4
Colorado	242	18	82	44	8	239	18	82	38	5
Connecticut	243	16	84	46	9	242	16	84	43	6
Delaware	242	13	87	40	5	241	13	87	40	4
Florida	243	13	87	43	7	241	14	86	38	5
Georgia	236	21	79	33	5	234	22	78	30	3
Hawaii	233	24	76	33	4	236	22	78	34	4
Idaho	242	16	84	42	6	240	15	85	38	5
Illinois	239	21	79	40	7	235	22	78	33	4
Indiana	246	11	89	48	7	244	12	88	45	6
Iowa	244	13	87	46	6	241	14	86	40	5
Kansas	249	11	89	54	10	247	10	90	48	8
Kentucky	237	19	81	33	4	234	22	78	29	3
Louisiana	230	28	72	25	3	230	27	73	24	2
Maine	244	14	86	43	7	241	15	85	40	5
Maryland	242	19	81	43	9	239	21	79	37	6
Massachusetts	254	7	93	60	13	251	7	93	55	9
Michigan	238	20	80	39	6	237	20	80	35	4
Minnesota	249	12	88	54	12	245	13	87	47	7
Mississippi	228	30	70	22	1	227	30	70	20	1
Missouri	240	17	83	40	6	238	19	81	37	4
Montana	245	12	88	47	6	242	13	87	42	4
Nebraska	240	18	82	40	6	236	22	78	35	4
Nevada	233	26	74	33	4	230	27	73	27	2
New Hampshire	250	8	92	54	8	247	10	90	49	7
New Jersey	250	10	90	55	11	247	11	89	49	8
New Mexico	229	29	71	26	2	227	30	70	23	2
New York	244	15	85	45	8	242	15	85	42	5
North Carolina	243	16	84	43	7	241	15	85	39	5
North Dakota	248	8	92	50	6	243	10	90	41	4
Ohio	246	11	89	49	8	243	14	86	43	5
Oklahoma	238	17	83	34	3	236	18	82	31	2
Oregon	238	20	80	38	6	234	23	77	32	3
Pennsylvania	245	15	85	50	9	243	15	85	44	5
Rhode Island	236	20	80	36	4	235	21	79	32	3
South Carolina	236	22	78	36	5	238	19	81	36	5
South Dakota	242	14	86	43	4	240	14	86	38	3
Tennessee	234	23	77	31	4	231	24	76	26	2
Texas	243	13	87	41	6	242	12	88	39	5
Utah	241	16	84	42	5	238	18	82	37	3
Vermont	248	11	89	51	9	245	11	89	47	6
Virginia	245	11	89	44	8	242	14	86	39	5
Washington	244	15	85	46	9	241	16	84	41	6
West Virginia	238	17	83	35	4	235	20	80	30	2
Wisconsin	245	15	85	48	8	243	15	85	46	6
Wyoming	244	12	88	46	5	243	11	89	43	4
Other jurisdictions										
District of Columbia	213	52	48	14	3	214	49	51	13	2
DoDEA[1]	241	13	87	39	4	239	15	85	35	2

Department of Defense Education Activity (overseas and domestic schools).

Note: Detail may not sum to totals because of rounding.

Source: U.S. Department of Education, Institute of Education Sciences, National Center for Education Statistics, National Assessment of Educational Progress (NAEP), 2007 Mathematics Assessment.

Table A-11. Average scale scores and achievement-level results in NAEP mathematics for fourth-grade public school students, by eligibility for free/ reduced-price school lunch and state: 2007

	Eligible					Not eligible					Information not available				
		Percentage of students					Percentage of students					Percentage of students			
State/jurisdiction	Average scale score	Below Basic	At or above Basic	At or above Proficient	At Advanced	Average scale score	Below Basic	At or above Basic	At or above Proficient	At Advanced	Average scale score	Below Basic	At or above Basic	At or above Proficient	At Advanced
Nation (public)	**227**	**30**	**70**	**22**	**1**	**249**	**9**	**91**	**53**	**9**	**243**	**17**	**83**	**44**	**8**
Alabama	217	43	57	13	1	242	14	86	41	5	‡	‡	‡	‡	‡
Alaska	225	34	66	23	2	247	11	89	50	9	‡	‡	‡	‡	‡
Arizona	219	40	60	15	1	245	12	88	46	7	255	6	94	64	11
Arkansas	229	27	73	24	1	249	9	91	54	8	‡	‡	‡	‡	‡
California	219	42	58	16	1	243	16	84	46	9	233	28	72	31	4
Colorado	225	33	67	21	2	251	8	92	55	9	‡	‡	‡	‡	‡
Connecticut	222	36	64	16	1	252	7	93	57	10	‡	‡	‡	‡	‡
Delaware	232	21	79	23	1	248	8	92	50	7	‡	‡	‡	‡	‡
Florida	233	21	79	25	2	251	7	93	55	9	‡	‡	‡	‡	‡
Georgia	224	32	68	16	1	247	9	91	49	7	‡	‡	‡	‡	‡
Hawaii	224	33	67	20	2	242	16	84	43	6	‡	‡	‡	‡	‡
Idaho	232	25	75	27	2	248	8	92	50	7	‡	‡	‡	‡	‡
Illinois	223	36	64	17	1	249	10	90	51	9	‡	‡	‡	‡	‡
Indiana	235	20	80	30	2	253	5	95	58	10	‡	‡	‡	‡	‡
Iowa	231	24	76	26	2	249	8	92	52	7	‡	‡	‡	‡	‡
Kansas	237	19	81	34	4	255	5	95	63	12	‡	‡	‡	‡	‡
Kentucky	226	30	70	18	1	245	10	90	46	6	‡	‡	‡	‡	‡
Louisiana	225	33	67	17	1	243	12	88	42	6	‡	‡	‡	‡	‡
Maine	232	23	77	27	2	248	10	90	51	8	‡	‡	‡	‡	‡
Maryland	225	36	64	19	2	248	12	88	51	11	‡	‡	‡	‡	‡
Massachusetts	237	17	83	32	3	258	3	97	67	14	‡	‡	‡	‡	‡
Michigan	224	35	65	20	1	246	11	89	48	7	‡	‡	‡	‡	‡
Minnesota	232	25	75	28	3	253	7	93	60	12	‡	‡	‡	‡	‡
Mississippi	222	38	62	13	#	241	13	87	39	3	240	14	86	40	3
Missouri	228	29	71	22	1	247	10	90	50	8	‡	‡	‡	‡	‡
Montana	234	22	78	30	2	250	6	94	54	7	‡	‡	‡	‡	‡
Nebraska	225	34	66	21	2	246	11	89	49	7	‡	‡	‡	‡	‡
Nevada	221	39	61	16	1	242	15	85	42	5	231	26	74	31	2
New Hampshire	236	18	82	32	2	251	7	93	57	9	‡	‡	‡	‡	‡
New Jersey	233	22	78	26	2	255	6	94	62	12	258	6	94	62	18
New Mexico	221	38	62	16	1	242	14	86	43	5	‡	‡	‡	‡	‡
New York	233	24	76	28	3	252	6	94	58	9	‡	‡	‡	‡	‡
North Carolina	231	24	76	24	2	252	7	93	57	10	238	18	82	40	2
North Dakota	235	18	82	30	2	250	5	95	53	6	‡	‡	‡	‡	‡
Ohio	230	25	75	23	1	253	5	95	59	9	‡	‡	‡	‡	‡
Oklahoma	230	25	75	22	1	245	9	91	46	5	‡	‡	‡	‡	‡
Oregon	226	32	68	21	1	245	12	88	47	7	231	23	77	27	3
Pennsylvania	227	29	71	22	1	253	7	93	61	10	‡	‡	‡	‡	‡
Rhode Island	222	35	65	18	1	245	11	89	45	5	‡	‡	‡	‡	‡
South Carolina	226	30	70	20	1	249	9	91	54	8	‡	‡	‡	‡	‡
South Dakota	230	25	75	25	1	247	8	92	49	5	‡	‡	‡	‡	‡
Tennessee	223	36	64	17	1	242	12	88	40	5	‡	‡	‡	‡	‡
Texas	235	18	82	27	2	252	6	94	56	9	255	5	95	62	12
Utah	229	29	71	25	2	246	11	89	48	6	‡	‡	‡	‡	‡
Vermont	234	20	80	31	2	252	7	93	57	10	‡	‡	‡	‡	‡
Virginia	230	24	76	20	1	250	8	92	52	9	‡	‡	‡	‡	‡
Washington	230	26	74	26	2	251	9	91	56	11	244	14	86	47	9
West Virginia	229	27	73	22	1	244	11	89	43	5	‡	‡	‡	‡	‡
Wisconsin	228	32	68	25	2	252	6	94	58	9	‡	‡	‡	‡	‡
Wyoming	236	18	82	32	2	248	8	92	51	6	‡	‡	‡	‡	‡
Other jurisdictions															
District of Columbia	207	57	43	7	#	228	36	64	27	7	‡	‡	‡	‡	‡
DoDEA[1]	‡	‡	‡	‡	‡	‡	‡	‡	‡	‡	240	14	86	37	3

Rounds to zero.

‡ Reporting standards not met. Sample size is insufficient to permit a reliable estimate.

[1] Department of Defense Education Activity (overseas and domestic schools).

Note: Detail may not sum to totals because of rounding.

Source: U.S. Department of Education, Institute of Education Sciences, National Center for Education Statistics, National Assessment of Educational Progress (NAEP), 2007 Mathematics Assessment.

Table A-12. Average scale scores and achievement-level results in NAEP mathematics for fourth-grade public school students, by status as students with disabilities (SD) and state: 2007

	SD					NotSD				
		Percentage of students					Percentage of students			
State/jurisdiction	Average scale	Below Basic	At or above Basic	At or above Proficient	At Advanced	Average scale	Below Basic	At or above Basic	At or above Proficient	At Advanced
Nation (public)	**220**	**40**	**60**	**19**	**2**	**241**	**16**	**84**	**41**	**6**
Alabama	197	69	31	8	1	232	25	75	28	3
Alaska	216	46	54	14	1	241	17	83	42	7
Arizona	209	54	46	13	2	234	24	76	32	4
Arkansas	216	49	51	18	2	240	16	84	39	5
California	205	59	41	14	2	232	28	72	31	5
Colorado	214	48	52	14	2	243	15	85	45	7
Connecticut	216	43	57	13	2	246	13	87	49	8
Delaware	227	32	68	22	2	244	10	90	43	5
Florida	223	37	63	18	1	245	10	90	44	6
Georgia	219	42	58	18	2	237	19	81	33	4
Hawaii	197	68	32	8	1	238	18	82	36	5
Idaho	216	47	53	14	1	243	12	88	43	6
Illinois	221	41	59	22	4	239	19	81	38	6
Indiana	228	28	72	25	2	248	8	92	50	7
Iowa	219	42	58	15	2	246	10	90	47	6
Kansas	226	35	65	23	3	251	8	92	54	9
Kentucky	223	37	63	19	2	237	18	82	33	4
Louisiana	213	52	48	11	1	233	22	78	27	2
Maine	226	32	68	21	2	245	11	89	46	7
Maryland	222	42	58	21	3	242	18	82	42	8
Massachusetts	238	17	83	33	4	255	5	95	61	12
Michigan	217	46	54	16	2	240	17	83	40	5
Minnesota	225	36	64	25	3	250	9	91	54	10
Mississippi	217	46	54	14	1	229	28	72	22	1
Missouri	225	35	65	23	2	241	16	84	40	6
Montana	223	38	62	18	1	246	9	91	47	6
Nebraska	220	40	60	17	2	241	16	84	41	5
Nevada	221	45	55	26	4	233	24	76	31	3
New Hampshire	230	25	75	25	1	252	5	95	57	9
New Jersey	229	30	70	25	3	251	8	92	56	10
New Mexico	208	56	44	9	#	230	27	73	26	2
New York	220	39	61	15	1	246	11	89	48	7
North Carolina	224	37	63	22	2	244	12	88	44	7
North Dakota	232	23	77	24	1	247	7	93	49	5
Ohio	227	29	71	22	2	247	10	90	49	7
Oklahoma	217	46	54	14	1	239	14	86	35	3
Oregon	216	46	54	16	1	239	18	82	38	5
Pennsylvania	223	38	62	26	3	248	11	89	51	8
Rhode Island	216	45	55	15	1	240	15	85	38	4
South Carolina	214	45	55	16	1	240	17	83	39	5
South Dakota	225	34	66	22	2	244	11	89	44	4
Tennessee	219	42	58	19	3	234	22	78	30	3
Texas	228	29	71	23	2	244	11	89	42	5
Utah	215	48	52	16	1	242	14	86	42	5
Vermont	221	39	61	16	1	251	6	94	55	8
Virginia	231	26	74	26	3	245	11	89	44	7
Washington	220	42	58	21	3	246	12	88	47	8
West Virginia	222	39	61	18	1	239	15	85	35	3
Wisconsin	223	37	63	21	2	247	12	88	51	8
Wyoming	224	36	64	19	1	247	8	92	48	5
Other jurisdictions										
District of Columbia	188	80	20	3	1	216	48	52	15	3
DoDEA[1]	218	43	57	13	#	243	11	89	40	3

Rounds to zero.

[1] Department of Defense Education Activity (overseas and domestic schools).

Note: The results for students with disabilities are based on students who were assessed and cannot be generalized to the total population of such students. Detail may not sum to totals because of rounding.

Source: U.S. Department of Education, Institute of Education Sciences, National Center for Education Statistics, National Assessment of Educational Progress (NAEP), 2007 Mathematics Assessment.

Table A-13. Average scale scores and achievement-level results in NAEP mathematics for fourth-grade public school students, by status as English language learners (ELL) and state: 2007

State/jurisdiction	ELL					Not ELL				
	Average scale	Percentage of students				Average scale	Percentage of students			
		Below *Basic*	At or above *Basic*	At or above *Proficient*	At *Advanced*		Below *Basic*	At or above *Basic*	At or above *Proficient*	At *Advanced*
Nation (public)	**217**	**44**	**56**	**13**	**1**	**242**	**16**	**84**	**42**	**6**
Alabama	213	51	49	11	2	229	29	71	26	3
Alaska	213	49	51	14	1	242	16	84	42	7
Arizona	203	64	36	6	1	237	20	80	35	5
Arkansas	222	35	65	16	#	239	18	82	38	5
California	212	51	49	10	1	239	20	80	40	6
Colorado	212	50	50	9	#	245	13	87	47	7
Connecticut	211	52	48	6	#	245	13	87	47	8
Delaware	226	27	73	14	#	242	13	87	41	5
Florida	223	36	64	16	1	243	12	88	42	6
Georgia	212	49	51	5	#	236	20	80	32	4
Hawaii	213	50	50	14	1	237	20	80	35	5
Idaho	214	51	49	10	#	243	12	88	43	6
Illinois	213	50	50	9	1	239	19	81	39	6
Indiana	233	23	77	26	3	246	10	90	47	7
Iowa	220	41	59	15	#	244	12	88	44	6
Kansas	229	28	72	21	2	250	9	91	54	9
Kentucky	221	38	62	16	1	235	20	80	31	3
Louisiana	‡	‡	‡	‡	‡	230	27	73	24	2
Maine	‡	‡	‡	‡	‡	243	14	86	42	6
Maryland	226	36	64	22	3	241	19	81	41	8
Massachusetts	230	26	74	24	2	254	6	94	60	11
Michigan	234	25	75	32	4	238	20	80	37	5
Minnesota	221	38	62	15	#	249	10	90	54	10
Mississippi	‡	‡	‡	‡	‡	228	30	70	21	1
Missouri	‡	‡	‡	‡	‡	240	18	82	39	5
Montana	215	47	53	6	#	245	11	89	46	5
Nebraska	211	52	48	8	#	240	17	83	40	5
Nevada	209	55	45	7	#	238	18	82	36	3
New Hampshire	229	31	69	25	2	249	8	92	52	8
New Jersey	218	45	55	14	1	250	9	91	53	10
New Mexico	209	55	45	7	#	233	23	77	29	3
New York	219	42	58	12	1	245	13	87	46	7
North Carolina	229	22	78	18	1	243	15	85	43	6
North Dakota	224	37	63	21	1	246	9	91	46	5
Ohio	231	29	71	27	5	245	12	88	46	7
Oklahoma	223	35	65	15	1	238	17	83	33	3
Oregon	210	56	44	7	#	240	17	83	39	5
Pennsylvania	211	53	47	8	2	245	14	86	48	7
Rhode Island	207	56	44	9	1	238	18	82	36	4
South Carolina	230	27	73	28	3	237	20	80	36	5
South Dakota	212	47	53	5	#	242	12	88	42	4
Tennessee	204	58	42	4	#	233	23	77	29	3
Texas	229	26	74	20	1	245	10	90	44	6
Utah	221	41	59	19	1	242	14	86	42	5
Vermont	230	31	69	28	6	247	11	89	50	7
Virginia	234	19	81	25	2	244	12	88	43	7
Washington	214	48	52	11	2	245	13	87	47	8
West Virginia	‡	‡	‡	‡	‡	236	19	81	32	3
Wisconsin	227	33	67	22	2	245	13	87	49	7
Wyoming	221	39	61	17	1	245	11	89	45	5
Other jurisdictions										
District of Columbia	209	58	42	9	1	214	50	50	14	3
DoDEA[1]	224	32	68	12	#	241	13	87	39	3

Rounds to zero.

‡ Reporting standards not met. Sample size is insufficient to permit a reliable estimate.

[1] Department of Defense Education Activity (overseas and domestic schools).

Note: The results for English language learners are based on students who were assessed and cannot be generalized to the total population of such students. Detail may not sum to totals because of rounding.

Source: U.S. Department of Education, Institute of Education Sciences, National Center for Education Statistics, National Assessment of Educational Progress (NAEP), 2007 Mathematics Assessment.

Table A-14. Percentage of eighth-grade public school students at or above *Basic* in NAEP mathematics, by state: Various years, 1990–2007

	Accommodations not permitted				Accommodations permitted			
State/jurisdiction	1990	1992	1996	2000	2000	2003	2005	2007
Nation (public)[1]	**51***	**56***	**61***	**65***	**62***	**67***	**68***	**70**
Alabama	40*	39*	45*	52	53	53	53	55
Alaska	—	—	68	—	—	70	69*	73
Arizona	48*	55*	57*	62	60*	61*	64	66
Arkansas	44*	44*	52*	52*	49*	58*	64	65
California	45*	50*	51*	52*	50*	56*	57	59
Colorado	57*	64*	67*	—	—	74	70*	75
Connecticut	60*	64*	70	72	70	73	70	73
Delaware	48*	52*	55*	—	—	68*	72	74
Florida	43*	49*	54*	—	—	62*	65	68
Georgia	47*	48*	51*	55*	54*	59*	62	64
Hawaii	40*	46*	51*	52*	51*	56*	56*	59
Idaho	63*	68*	—	71	70*	73	73	75
Illinois	50*	—	—	68	67	66	68	70
Indiana	56*	60*	68*	76	74	74	74	76
Iowa	70*	76	78	—	—	76	75	77
Kansas	—	—	—	77	76*	76*	77*	81
Kentucky	43*	51*	56*	63*	60*	65	64*	69
Louisiana	32*	37*	38*	48*	47*	57*	59	64
Maine	—	72*	77	76	73*	75*	74*	78
Maryland	50*	54*	57*	65*	62*	67*	66*	74
Massachusetts	—	63*	68*	76*	70*	76*	80*	85
Michigan	53*	58*	67	70	68	68	68	66
Minnesota	67*	74*	75*	80	80	82	79	81
Mississippi	—	33*	36*	41*	42*	47*	52	54
Missouri	—	62*	64*	67*	64*	71	68	72
Montana	74*	—	75	80	79	79	80	79
Nebraska	68*	70*	76	74	73	74	75	74
Nevada	—	—	—	58	55*	59	60	60
New Hampshire	65*	71*	—	—	—	79	77	78
New Jersey	58*	62*	—	—	—	72*	74	77
New Mexico	43*	48*	51*	50*	48*	52*	53	57
New York	50*	57*	61*	68	63*	70	70	70
North Carolina	38*	47*	56*	70	67*	72	72	73
North Dakota	75*	78*	77*	77*	76*	81*	81*	86
Ohio	53*	59*	—	75	73	74	74	76
Oklahoma	52*	59*	—	64	62	65	63	66
Oregon	62*	—	67*	71	71	70	72	73
Pennsylvania	56*	62*	—	—	—	69*	72*	77
Rhode Island	49*	56*	60*	64	59*	63	63	65
South Carolina	—	48*	48*	55*	53*	68	71	71
South Dakota	—	—	—	—	—	78	80	81
Tennessee	—	47*	53*	53*	52*	59	61	64
Texas	45*	53*	59*	68*	67*	69*	72*	78
Utah	—	67*	70	68*	66*	72	71	72
Vermont	—	—	72*	75*	73*	77*	78*	81
Virginia	52*	57*	58*	67*	65*	72*	75	77
Washington	—	—	67*	—	—	72	75	75
West Virginia	42*	47*	54*	62	58	63	60	61
Wisconsin	66*	71*	75	—	—	75	76	76
Wyoming	64*	67*	68*	70*	69*	77*	76*	80
Other jurisdictions								
District of Columbia	17*	22*	20*	23*	23*	29*	31	34
DoDEA[2]	—	—	64*	70*	68*	79	76	78

— Not available. The jurisdiction did not participate or did not meet the minimum participation guidelines for reporting.

* Significantly different ($p < .05$) from 2007 when only one jurisdiction or the nation is being examined.

[1] National results for assessments prior to 2003 are based on the national sample, not on aggregated state samples.

[2] Department of Defense Education Activity (overseas and domestic schools). Before 2005, DoDEA overseas and domestic schools were separate jurisdictions in NAEP. Pre-2005 data presented here were recalculated for comparability.

Source: U.S. Department of Education, Institute of Education Sciences, National Center for Education Statistics, National Assessment of Educational Progress (NAEP), various years, 1990–2007 Mathematics Assessments.

Table A-15. Percentage of eighth-grade public school students at or above *Proficient* in NAEP mathematics, by state: Various years, 1990–2007

State/jurisdiction	Accommodations not permitted				Accommodations permitted			
	1990	1992	1996	2000	2000	2003	2005	2007
Nation (public)[1]	**15***	**20***	**23***	**26***	**25***	**27***	**28***	**31**
Alabama	9*	10*	12*	16	16	16	15	18
Alaska	—	—	30	—	—	30	29	32
Arizona	13*	15*	18*	21*	20*	21*	26	26
Arkansas	9*	10*	13*	14*	13*	19*	22	24
California	12*	16*	17*	18*	17*	22	22*	24
Colorado	17*	22*	25*	—	—	34	32*	37
Connecticut	22*	26*	31	34	33	35	35	35
Delaware	14*	15*	19*	—	—	26*	30	31
Florida	12*	15*	17*	—	—	23*	26	27
Georgia	14*	13*	16*	19*	19*	22*	23	25
Hawaii	12*	14*	16*	16*	16*	17*	18*	21
Idaho	18*	22*	—	27*	26*	28*	30*	34
Illinois	15*	—	—	27	26*	29	29	31
Indiana	17*	20*	24*	31	29*	31*	30*	35
Iowa	25*	31*	31	—	—	33	34	35
Kansas	—	—	—	34*	34*	34*	34*	40
Kentucky	10*	14*	16*	21*	20*	24*	23*	27
Louisiana	5*	7*	7*	12*	11*	17	16	19
Maine	—	25*	31	32	30	29*	30*	34
Maryland	17*	20*	24*	29*	27*	30*	30*	37
Massachusetts	—	23*	28*	32*	30*	38*	43*	51
Michigan	16*	19*	28	28	28	28	29	29
Minnesota	23*	31*	34*	40	39	44	43	43
Mississippi	—	6*	7*	8*	9*	12	14	14
Missouri	—	20*	22*	22*	21*	28	26*	30
Montana	27*	—	32*	37	36	35	36	38
Nebraska	24*	26*	31	31	30*	32	35	35
Nevada	—	—	—	20*	18*	20*	21	23
New Hampshire	20*	25*	—	—	—	35	35	38
New Jersey	21*	24*	—	—	—	33*	36*	40
New Mexico	10*	11*	14	13*	12*	15	14*	17
New York	15*	20*	22*	26	24*	32	31	30
North Carolina	9*	12*	20*	30*	27*	32	32	34
North Dakota	27*	29*	33*	31*	30*	36*	35*	41
Ohio	15*	18*	—	31*	30*	30*	33	35
Oklahoma	13*	17*	—	19	18	20	21	21
Oregon	21*	—	26*	32	31	32	34	35
Pennsylvania	17*	21*	—	—	—	30*	31*	38
Rhode Island	15*	16*	20*	24*	22*	24*	24*	28
South Carolina	—	15*	14*	18*	17*	26*	30	32
South Dakota	—	—	—	—	—	35*	36	39
Tennessee	—	12*	15*	17*	16*	21	21	23
Texas	13*	18*	21*	24*	24*	25*	31*	35
Utah	—	22*	24*	26*	25*	31	30	32
Vermont	—	—	27*	32*	31*	35*	38*	41
Virginia	17*	19*	21*	26*	25*	31*	33	37
Washington	—	—	26*	—	—	32*	36	36
West Virginia	9*	10*	14*	18	17	20	18	19
Wisconsin	23*	27*	32*	—	—	35	36	37
Wyoming	19*	21*	22*	25*	23*	32	29*	36
Other jurisdictions								
District of Columbia	3*	4*	5*	6*	6*	6*	7	8
DoDEA[2]	—	—	22*	27*	26*	33	33	33

— Not available. The jurisdiction did not participate or did not meet the minimum participation guidelines for reporting. * Significantly different ($p < .05$) from 2007 when only one jurisdiction or the nation is being examined.

[1] National results for assessments prior to 2003 are based on the national sample, not on aggregated state samples.

[2] Department of Defense Education Activity (overseas and domestic schools). Before 2005, DoDEA overseas and domestic schools were separate jurisdictions in NAEP. Pre-2005 data presented here were

recalculated for comparability.

Source: U.S. Department of Education, Institute of Education Sciences, National Center for Education Statistics, National Assessment of Educational Progress (NAEP), various years, 1990–2007 Mathematics Assessments.

Table A-16. Average scale scores and achievement-level results in NAEP mathematics for eighth-grade public school students, by race/ethnicity and state: 2007

State/jurisdiction	White					Black					Hispanic				
		Percentage of students					Percentage of students					Percentage of students			
	Average scale score	Below *Basic*	At or above *Basic*	At or above *Proficient*	At *Advanced*	Average scale score	Below *Basic*	At or above *Basic*	At or above *Proficient*	At *Advanced*	Average scale score	Below *Basic*	At or above *Basic*	At or above *Proficient*	At *Advanced*
Nation (public)	**290**	**19**	**81**	**41**	**9**	**259**	**53**	**47**	**11**	**1**	**264**	**46**	**54**	**15**	**2**
Alabama	278	30	70	27	4	246	69	31	4	#	249	63	37	3	#
Alaska	294	14	86	44	10	271	37	63	15	3	274	34	66	23	2
Arizona	289	19	81	40	8	266	42	58	15	2	262	48	52	12	1
Arkansas	282	26	74	31	5	254	58	42	9	1	256	54	46	8	#
California	287	22	78	39	8	253	62	38	10	1	256	56	44	10	1
Colorado	296	15	85	48	13	272	40	60	21	4	264	47	53	13	2
Connecticut	293	17	83	44	11	255	56	44	7	#	254	56	44	10	1
Delaware	294	14	86	43	9	265	44	56	10	1	267	42	58	17	1
Florida	289	20	80	37	8	259	52	48	11	1	270	39	61	21	3
Georgia	288	20	80	37	6	261	52	48	11	1	266	45	55	16	2
Hawaii	278	28	72	28	5	‡	‡	‡	‡	‡	264	47	53	15	1
Idaho	287	21	79	38	7	‡	‡	‡	‡	‡	264	47	53	16	2
Illinois	291	19	81	41	9	253	59	41	7	#	265	45	55	13	1
Indiana	290	18	82	40	9	259	53	47	9	#	267	45	55	20	2
Iowa	288	19	81	38	7	257	60	40	11	3	261	50	50	13	1
Kansas	295	13	87	46	10	267	43	57	16	2	269	42	58	16	2
Kentucky	282	27	73	29	5	257	58	42	11	1	‡	‡	‡	‡	‡
Louisiana	283	21	79	28	3	258	56	44	7	1	‡	‡	‡	‡	‡
Maine	287	21	79	35	7	‡	‡	‡	‡	‡	‡	‡	‡	‡	‡
Maryland	300	12	88	53	15	265	47	53	13	1	272	36	64	21	3
Massachusetts	305	9	91	58	17	264	46	54	13	1	270	41	59	19	5
Michigan	285	24	76	35	8	244	72	28	5	#	259	56	44	11	#
Minnesota	297	14	86	48	13	260	52	48	14	1	269	44	56	18	2
Mississippi	279	26	74	24	3	251	65	35	4	#	‡	‡	‡	‡	‡
Missouri	288	19	81	36	7	253	62	38	6	#	270	38	62	17	1
Montana	291	17	83	41	8	‡	‡	‡	‡	‡	‡	‡	‡	‡	‡
Nebraska	291	18	82	41	9	240	72	28	5	1	261	50	50	11	2
Nevada	282	27	73	32	5	255	56	44	12	1	257	56	44	11	1
New Hampshire	289	21	79	39	8	‡	‡	‡	‡	‡	264	46	54	14	2
New Jersey	298	13	87	51	14	264	45	55	14	1	271	37	63	20	2
New Mexico	285	23	77	33	6	264	48	52	12	2	260	52	48	10	1
New York	290	18	82	39	8	258	54	46	10	1	264	46	54	15	2
North Carolina	295	15	85	46	12	266	47	53	14	1	273	39	61	23	4
North Dakota	295	11	89	44	7	‡	‡	‡	‡	‡	‡	‡	‡	‡	‡
Ohio	291	17	83	42	8	258	53	47	9	#	276	37	63	25	5
Oklahoma	280	26	74	25	4	258	57	43	9	1	259	54	46	8	#
Oregon	289	22	78	39	10	272	41	59	28	3	261	50	50	14	1
Pennsylvania	293	16	84	44	9	257	55	45	13	1	264	45	55	17	3
Rhode Island	284	25	75	35	6	250	61	39	9	#	251	61	39	7	1
South Carolina	293	17	83	44	11	265	45	55	15	1	272	38	62	23	5
South Dakota	292	15	85	43	8	‡	‡	‡	‡	‡	269	43	57	18	5
Tennessee	282	25	75	30	5	254	62	38	7	1	264	49	51	13	2
Texas	300	10	90	53	13	271	36	64	16	1	277	30	70	23	3
Utah	286	22	78	36	7	‡	‡	‡	‡	‡	256	56	44	12	1
Vermont	292	18	82	42	10	‡	‡	‡	‡	‡	‡	‡	‡	‡	‡
Virginia	296	14	86	47	12	268	44	56	15	1	275	36	64	24	5
Washington	291	19	81	42	10	264	44	56	16	4	263	46	54	13	2
West Virginia	271	37	63	19	2	250	69	31	4	#	‡	‡	‡	‡	‡
Wisconsin	292	17	83	42	9	247	70	30	6	#	268	41	59	18	2
Wyoming	290	17	83	39	7	‡	‡	‡	‡	‡	274	36	64	22	3
Other jurisdictions															
District of Columbia	‡	‡	‡	‡	‡	245	69	31	6	#	251	62	38	9	1
DoDEA[1]	291	16	84	40	7	272	36	64	15	2	282	26	74	28	4

See notes at end of table.

Table A-16. Continued

State/jurisdiction	Asian/Pacific Islander: Average scale score	Asian/Pacific Islander: Below *Basic*	Asian/Pacific Islander: At or above *Basic*	Asian/Pacific Islander: At or above *Proficient*	Asian/Pacific Islander: At *Advanced*	American Indian/Alaska Native: Average scale score	American Indian/Alaska Native: Below *Basic*	American Indian/Alaska Native: At or above *Basic*	American Indian/Alaska Native: At or above *Proficient*	American Indian/Alaska Native: At *Advanced*
		Percentage of students					Percentage of students			
Nation (public)	**296**	**18**	**82**	**49**	**17**	**265**	**44**	**56**	**17**	**2**
Alabama	‡	‡	‡	‡	‡	‡	‡	‡	‡	‡
Alaska	282	29	71	33	6	260	51	49	12	2
Arizona	303	11	89	52	22	258	50	50	12	1
Arkansas	‡	‡	‡	‡	‡	‡	‡	‡	‡	‡
California	293	21	79	46	14	263	50	50	17	3
Colorado	297	18	82	48	17	‡	‡	‡	‡	‡
Connecticut	307	8	92	61	24	‡	‡	‡	‡	‡
Delaware	309	11	89	65	26	‡	‡	‡	‡	‡
Florida	293	20	80	48	14	‡	‡	‡	‡	‡
Georgia	‡	‡	‡	‡	‡	‡	‡	‡	‡	‡
Hawaii	268	42	58	20	3	‡	‡	‡	‡	‡
Idaho	‡	‡	‡	‡	‡	‡	‡	‡	‡	‡
Illinois	303	13	87	55	23	‡	‡	‡	‡	‡
Indiana	‡	‡	‡	‡	‡	‡	‡	‡	‡	‡
Iowa	‡	‡	‡	‡	‡	‡	‡	‡	‡	‡
Kansas	302	14	86	52	23	‡	‡	‡	‡	‡
Kentucky	‡	‡	‡	‡	‡	‡	‡	‡	‡	‡
Louisiana	‡	‡	‡	‡	‡	‡	‡	‡	‡	‡
Maine	‡	‡	‡	‡	‡	‡	‡	‡	‡	‡
Maryland	313	8	92	66	30	‡	‡	‡	‡	‡
Massachusetts	315	6	94	74	28	‡	‡	‡	‡	‡
Michigan	‡	‡	‡	‡	‡	‡	‡	‡	‡	‡
Minnesota	283	28	72	34	8	266	43	57	19	2
Mississippi	‡	‡	‡	‡	‡	‡	‡	‡	‡	‡
Missouri	‡	‡	‡	‡	‡	‡	‡	‡	‡	‡
Montana	‡	‡	‡	‡	‡	260	50	50	15	2
Nebraska	‡	‡	‡	‡	‡	‡	‡	‡	‡	‡
Nevada	285	24	76	36	7	‡	‡	‡	‡	‡
New Hampshire	‡	‡	‡	‡	‡	‡	‡	‡	‡	‡
New Jersey	314	7	93	69	30	‡	‡	‡	‡	‡
New Mexico	‡	‡	‡	‡	‡	253	60	40	7	1
New York	302	14	86	53	23	‡	‡	‡	‡	‡
North Carolina	299	15	85	50	18	261	49	51	17	1
North Dakota	‡	‡	‡	‡	‡	264	44	56	14	1
Ohio	‡	‡	‡	‡	‡	‡	‡	‡	‡	‡
Oklahoma	‡	‡	‡	‡	‡	269	40	60	17	2
Oregon	299	18	82	53	17	264	51	49	16	3
Pennsylvania	314	9	91	66	36	‡	‡	‡	‡	‡
Rhode Island	282	29	71	31	8	‡	‡	‡	‡	‡
South Carolina	‡	‡	‡	‡	‡	‡	‡	‡	‡	‡
South Dakota	‡	‡	‡	‡	‡	261	46	54	14	1
Tennessee	‡	‡	‡	‡	‡	‡	‡	‡	‡	‡
Texas	309	8	92	67	21	‡	‡	‡	‡	‡
Utah	277	32	68	32	5	‡	‡	‡	‡	‡
Vermont	‡	‡	‡	‡	‡	‡	‡	‡	‡	‡
Virginia	299	16	84	53	18	‡	‡	‡	‡	‡
Washington	289	24	76	41	14	265	45	55	18	3
West Virginia	‡	‡	‡	‡	‡	‡	‡	‡	‡	‡
Wisconsin	290	23	77	40	12	‡	‡	‡	‡	‡
Wyoming	‡	‡	‡	‡	‡	‡	‡	‡	‡	‡
Other jurisdictions										
District of Columbia	‡	‡	‡	‡	‡	‡	‡	‡	‡	‡
DoDEA[1]	284	23	77	34	5	‡	‡	‡	‡	‡

Rounds to zero.

‡ Reporting standards not met. Sample size is insufficient to permit a reliable estimate.

[1] Department of Defense Education Activity (overseas and domestic schools).

Note: Black includes African American, Hispanic includes Latino, and Pacific Islander includes Native Hawaiian. Race categories exclude Hispanic origin. Results are not shown for students whose race/ethnicity was "unclassified." Detail may not sum to totals because of rounding.

Source: U.S. Department of Education, Institute of Education Sciences, National Center for Education Statistics, National Assessment of Educational Progress (NAEP), 2007 Mathematics Assessment.

Table A-17. Average scale scores and achievement-level results in NAEP mathematics for eighth-grade public school students, by gender and state: 2007

State/jurisdiction	Male					Female				
		Percentage of students					Percentage of students			
	Average scale	Below *Basic*	At or above *Basic*	At or above *Proficient*	At *Advanced*	Average scale	Below *Basic*	At or above *Basic*	At or above *Proficient*	At *Advanced*
Nation (public)	**281**	**29**	**71**	**33**	**8**	**279**	**30**	**70**	**29**	**6**
Alabama	267	44	56	21	3	265	45	55	15	2
Alaska	282	27	73	33	8	283	27	73	32	6
Arizona	277	32	68	30	6	274	35	65	23	4
Arkansas	274	37	63	26	5	274	34	66	22	3
California	270	41	59	25	6	270	40	60	23	4
Colorado	287	25	75	38	10	286	25	75	37	9
Connecticut	282	29	71	35	9	283	25	75	34	8
Delaware	285	24	76	34	8	281	27	73	29	5
Florida	278	32	68	29	6	277	32	68	26	5
Georgia	275	36	64	26	5	274	36	64	23	4
Hawaii	267	42	58	20	4	270	39	61	22	2
Idaho	285	24	76	36	7	282	26	74	32	5
Illinois	282	29	71	33	8	279	31	69	29	6
Indiana	286	24	76	37	9	284	25	75	33	6
Iowa	287	21	79	37	8	284	24	76	33	6
Kansas	291	19	81	41	10	289	18	82	39	7
Kentucky	280	30	70	30	6	277	32	68	24	4
Louisiana	273	36	64	20	3	272	37	63	18	2
Maine	288	21	79	37	7	285	23	77	32	7
Maryland	287	25	75	38	12	284	27	73	35	9
Massachusetts	300	14	86	53	17	296	16	84	48	13
Michigan	278	32	68	30	7	275	35	65	27	5
Minnesota	292	19	81	44	12	292	19	81	43	11
Mississippi	266	44	56	16	2	264	48	52	12	1
Missouri	282	27	73	32	7	279	28	72	28	4
Montana	287	22	78	39	8	287	20	80	36	7
Nebraska	285	24	76	37	8	282	27	73	32	7
Nevada	271	39	61	24	4	270	41	59	22	3
New Hampshire	288	22	78	38	8	287	23	77	38	7
New Jersey	290	23	77	43	12	288	22	78	38	9
New Mexico	268	43	57	19	3	267	44	56	16	2
New York	281	30	70	31	8	280	29	71	29	6
North Carolina	285	26	74	36	9	283	28	72	33	7
North Dakota	293	14	86	43	8	290	15	85	39	6
Ohio	286	23	77	38	8	283	24	76	33	5
Oklahoma	277	32	68	24	4	273	35	65	18	2
Oregon	285	27	73	37	10	283	27	73	33	7
Pennsylvania	289	21	79	42	10	283	25	75	35	6
Rhode Island	276	34	66	29	6	275	35	65	27	4
South Carolina	281	29	71	33	8	282	29	71	31	7
South Dakota	290	19	81	41	8	287	19	81	37	5
Tennessee	277	34	66	26	5	271	38	62	20	3
Texas	287	22	78	37	8	285	23	77	32	6
Utah	282	27	73	34	7	280	29	71	30	5
Vermont	292	19	81	43	12	290	19	81	40	9
Virginia	289	22	78	40	10	286	24	76	34	8
Washington	285	26	74	37	10	285	24	76	35	8
West Virginia	271	38	62	21	3	269	40	60	16	2
Wisconsin	287	24	76	40	10	284	24	76	34	6
Wyoming	288	20	80	37	7	286	20	80	34	6
Other jurisdictions										
District of Columbia	248	66	34	8	1	248	66	34	8	1
DoDEA[1]	285	23	77	34	6	285	22	78	32	4

[1] Department of Defense Education Activity (overseas and domestic schools).

Note: Detail may not sum to totals because of rounding.

Source: U.S. Department of Education, Institute of Education Sciences, National Center for Education Statistics, National Assessment of Educational Progress (NAEP), 2007 Mathematics Assessment.

Table A-18. Average scale scores and achievement-level results in NAEP mathematics for eighth-grade public school students, by eligibility for free/ reduced-price school lunch and state: 2007

State/jurisdiction	Eligible					Not eligible					Information not available				
	Average scale score	Percentage of students				Average scale score	Percentage of students				Average scale score	Percentage of students			
		Below Basic	At or above Basic	At or above Proficient	At Advanced		Below Basic	At or above Basic	At or above Proficient	At Advanced		Below Basic	At or above Basic	At or above Proficient	At Advanced
Nation (public)	**265**	**45**	**55**	**15**	**2**	**291**	**19**	**81**	**42**	**10**	**274**	**36**	**64**	**28**	**6**
Alabama	250	63	37	6	#	281	27	73	30	5	‡	‡	‡	‡	‡
Alaska	266	45	55	17	3	292	16	84	41	10	‡	‡	‡	‡	‡
Arizona	262	48	52	13	1	286	23	77	36	8	294	18	82	48	8
Arkansas	263	46	54	14	2	285	23	77	35	6	‡	‡	‡	‡	‡
California	257	54	46	12	1	283	28	72	36	9	266	43	57	24	5
Colorado	267	42	58	17	2	296	16	84	48	14	‡	‡	‡	‡	‡
Connecticut	256	53	47	10	1	292	18	82	44	11	‡	‡	‡	‡	‡
Delaware	270	39	61	16	2	290	19	81	39	9	‡	‡	‡	‡	‡
Florida	265	45	55	16	1	287	22	78	37	9	‡	‡	‡	‡	‡
Georgia	262	51	49	12	1	287	22	78	36	7	‡	‡	‡	‡	‡
Hawaii	258	52	48	13	1	276	33	67	27	4	‡	‡	‡	‡	‡
Idaho	273	36	64	22	3	290	19	81	41	8	‡	‡	‡	‡	‡
Illinois	262	49	51	13	2	292	17	83	42	10	‡	‡	‡	‡	‡
Indiana	271	39	61	20	3	293	16	84	43	10	‡	‡	‡	‡	‡
Iowa	270	39	61	20	3	292	16	84	42	9	‡	‡	‡	‡	‡
Kansas	275	33	67	23	3	299	11	89	50	12	‡	‡	‡	‡	‡
Kentucky	267	43	57	15	1	288	21	79	37	8	‡	‡	‡	‡	‡
Louisiana	264	47	53	11	1	284	21	79	30	4	‡	‡	‡	‡	‡
Maine	275	33	67	21	3	292	16	84	40	9	‡	‡	‡	‡	‡
Maryland	268	43	57	15	3	293	20	80	45	13	‡	‡	‡	‡	‡
Massachusetts	275	35	65	25	4	306	8	92	60	19	‡	‡	‡	‡	‡
Michigan	259	53	47	14	1	285	24	76	36	8	‡	‡	‡	‡	‡
Minnesota	273	36	64	22	3	298	13	87	50	14	‡	‡	‡	‡	‡
Mississippi	257	57	43	7	#	280	25	75	26	3	‡	‡	‡	‡	‡
Missouri	266	45	55	16	2	290	16	84	39	8	‡	‡	‡	‡	‡
Montana	272	36	64	22	2	295	13	87	46	10	‡	‡	‡	‡	‡
Nebraska	265	45	55	17	2	293	16	84	43	10	‡	‡	‡	‡	‡
Nevada	259	53	47	13	2	279	31	69	30	5	265	44	56	16	1
New Hampshire	271	40	60	18	3	291	19	81	42	9	291	19	81	38	11
New Jersey	266	43	57	17	2	297	14	86	50	14	‡	‡	‡	‡	‡
New Mexico	258	55	45	9	1	282	27	73	30	6	‡	‡	‡	‡	‡
New York	268	43	57	19	4	292	16	84	42	9	‡	‡	‡	‡	‡
North Carolina	268	42	58	17	2	296	15	85	48	13	‡	‡	‡	‡	‡
North Dakota	280	27	73	29	4	296	10	90	45	8	‡	‡	‡	‡	‡
Ohio	268	40	60	16	1	293	16	84	44	9	‡	‡	‡	‡	‡
Oklahoma	264	46	54	13	1	285	21	79	30	6	‡	‡	‡	‡	‡
Oregon	270	41	59	20	3	294	17	83	45	13	275	35	65	26	4
Pennsylvania	267	41	59	19	2	294	16	84	46	10	‡	‡	‡	‡	‡
Rhode Island	257	55	45	10	1	285	24	76	36	7	‡	‡	‡	‡	‡
South Carolina	269	41	59	18	2	294	17	83	45	12	‡	‡	‡	‡	‡
South Dakota	275	31	69	24	3	294	13	87	46	9	‡	‡	‡	‡	‡
Tennessee	262	50	50	12	1	284	24	76	32	6	‡	‡	‡	‡	‡
Texas	275	32	68	21	2	297	12	88	49	12	‡	‡	‡	‡	‡
Utah	267	42	58	19	3	287	22	78	38	7	‡	‡	‡	‡	‡
Vermont	277	31	69	24	3	296	14	86	48	13	‡	‡	‡	‡	‡
Virginia	268	43	57	15	2	295	16	84	46	12	‡	‡	‡	‡	‡
Washington	268	41	59	19	3	294	17	83	45	12	‡	‡	‡	‡	‡
West Virginia	260	51	49	10	1	279	27	73	26	4	‡	‡	‡	‡	‡
Wisconsin	266	44	56	18	2	293	16	84	45	11	‡	‡	‡	‡	‡
Wyoming	275	33	67	23	3	291	15	85	41	8	‡	‡	‡	‡	‡
Other jurisdictions															
District of Columbia	243	72	28	4	#	259	55	45	15	2	‡	‡	‡	‡	‡
DoDEA[1]	‡	‡	‡	‡	‡	‡	‡	‡	‡	‡	285	22	78	33	5

Rounds to zero.

‡ Reporting standards not met. Sample size is insufficient to permit a reliable estimate.

[1] Department of Defense Education Activity (overseas and domestic schools).

Note: Detail may not sum to totals because of rounding.

Source: U.S. Department of Education, Institute of Education Sciences, National Center for Education Statistics, National Assessment of Educational Progress (NAEP), 2007 Mathematics Assessment.

Table A-19. Average scale scores and achievement-level results in NAEP mathematics for eighth-grade public school students, by status as students with disabilities (SD) and state: 2007

State/jurisdiction	SD					Not SD				
		Percentage of students					Percentage of students			
	Average scale score	Below *Basic*	At or above *Basic*	At or above *Proficient*	At *Advanced*	Average scale score	Below *Basic*	At or above *Basic*	At or above *Proficient*	At *Advanced*
Nation (public)	**246**	**67**	**33**	**8**	**1**	**284**	**26**	**74**	**33**	**7**
Alabama	220	91	9	1	#	271	40	60	20	3
Alaska	245	71	29	7	1	286	23	77	35	8
Arizona	237	73	27	4	#	279	30	70	28	5
Arkansas	233	82	18	3	1	279	30	70	27	4
California	228	81	19	5	1	274	38	62	25	5
Colorado	254	60	40	11	3	289	21	79	40	10
Connecticut	245	63	37	9	1	287	22	78	38	9
Delaware	258	56	44	12	2	285	23	77	33	7
Florida	246	66	34	8	1	281	27	73	30	6
Georgia	246	66	34	6	1	276	34	66	26	4
Hawaii	224	85	15	2	#	275	35	65	24	3
Idaho	245	71	29	5	1	287	21	79	37	7
Illinois	246	68	32	7	#	284	26	74	33	8
Indiana	254	60	40	11	1	289	20	80	38	8
Iowa	247	67	33	6	1	291	16	84	40	8
Kansas	257	57	43	9	2	293	15	85	43	9
Kentucky	249	65	35	7	#	281	28	72	29	5
Louisiana	242	73	27	4	#	276	32	68	21	2
Maine	259	54	46	11	1	290	17	83	37	8
Maryland	262	51	49	16	4	287	25	75	38	10
Massachusetts	271	38	62	18	2	301	13	87	54	16
Michigan	238	76	24	4	#	281	29	71	32	7
Minnesota	256	58	42	11	1	296	15	85	47	13
Mississippi	230	86	14	#	#	268	43	57	15	2
Missouri	249	64	36	7	1	284	24	76	32	6
Montana	248	67	33	5	1	292	16	84	41	8
Nebraska	248	64	36	8	1	288	21	79	38	8
Nevada	240	72	28	9	2	274	37	63	24	4
New Hampshire	258	56	44	9	1	293	16	84	44	9
New Jersey	251	62	38	9	1	294	17	83	45	12
New Mexico	240	77	23	6	1	271	40	60	19	3
New York	249	64	36	6	#	284	25	75	33	7
North Carolina	257	57	43	14	2	287	23	77	37	9
North Dakota	263	46	54	9	1	294	12	88	44	7
Ohio	250	63	37	7	1	288	20	80	38	7
Oklahoma	242	75	25	3	#	277	31	69	23	3
Oregon	251	63	37	9	2	287	23	77	37	9
Pennsylvania	254	56	44	14	2	291	19	81	42	9
Rhode Island	243	71	29	5	#	281	28	72	32	6
South Carolina	245	68	32	7	#	285	26	74	34	8
South Dakota	251	62	38	8	1	292	15	85	42	7
Tennessee	246	68	32	15	2	276	34	66	24	4
Texas	250	64	36	8	1	288	19	81	37	7
Utah	234	79	21	3	1	285	24	76	35	7
Vermont	261	52	48	12	2	296	13	87	47	12
Virginia	260	55	45	13	2	290	20	80	40	10
Washington	240	72	28	7	1	289	21	79	38	10
West Virginia	237	79	21	4	#	276	32	68	21	3
Wisconsin	249	63	37	8	#	290	19	81	40	9
Wyoming	252	65	35	6	#	292	14	86	40	7
Other jurisdictions										
District of Columbia	211	93	7	1	#	252	63	37	9	1
DoDEA[1]	252	65	35	6	2	288	19	81	35	5

Rounds to zero.

[1] Department of Defense Education Activity (overseas and domestic schools).

Note: The results for students with disabilities are based on students who were assessed and cannot be generalized to the total population of such students. Detail may not sum to totals because of rounding.

Source: U.S. Department of Education, Institute of Education Sciences, National Center for Education Statistics, National Assessment of Educational Progress (NAEP), 2007 Mathematics Assessment.

Table A-20. Average scale scores and achievement-level results in NAEP mathematics for eighth-grade public school students, by status as English language learners (ELL) and state: 2007

State/jurisdiction	ELL					Not ELL				
		Percentage of students					Percentage of students			
	Average scale	Below *Basic*	At or above *Basic*	At or above *Proficient*	At *Advanced*	Average scale	Below *Basic*	At or above *Basic*	At or above *Proficient*	At *Advanced*
Nation (public)	**245**	**70**	**30**	**6**	**1**	**282**	**27**	**73**	**33**	**7**
Alabama	‡	‡	‡	‡	‡	266	44	56	18	2
Alaska	254	59	41	8	1	288	21	79	37	8
Arizona	238	76	24	4	1	279	29	71	29	5
Arkansas	247	69	31	4	1	275	34	66	25	4
California	241	74	26	5	1	278	32	68	29	6
Colorado	244	72	28	3	1	289	22	78	40	10
Connecticut	227	87	13	1	#	285	25	75	36	9
Delaware	‡	‡	‡	‡	‡	284	25	75	32	7
Florida	243	72	28	6	1	279	30	70	28	6
Georgia	237	80	20	1	#	276	35	65	25	4
Hawaii	233	82	18	3	1	271	38	62	22	3
Idaho	247	70	30	7	#	286	23	77	36	7
Illinois	257	56	44	12	3	281	29	71	31	7
Indiana	261	55	45	17	4	286	23	77	36	8
Iowa	253	59	41	7	1	286	22	78	36	7
Kansas	255	58	42	8	#	292	17	83	42	9
Kentucky	‡	‡	‡	‡	‡	279	31	69	28	5
Louisiana	‡	‡	‡	‡	‡	272	36	64	19	2
Maine	‡	‡	‡	‡	‡	287	21	79	34	7
Maryland	‡	‡	‡	‡	‡	286	26	74	37	10
Massachusetts	251	67	33	16	3	299	13	87	52	15
Michigan	‡	‡	‡	‡	‡	277	33	67	29	6
Minnesota	258	54	46	12	1	293	17	83	45	12
Mississippi	‡	‡	‡	‡	‡	265	46	54	14	2
Missouri	‡	‡	‡	‡	‡	281	27	73	30	5
Montana	237	75	25	1	#	289	18	82	39	8
Nebraska	241	77	23	1	#	285	24	76	35	8
Nevada	238	77	23	5	#	274	36	64	25	4
New Hampshire	‡	‡	‡	‡	‡	288	22	78	38	8
New Jersey	257	55	45	11	#	290	21	79	41	11
New Mexico	242	75	25	3	#	272	38	62	20	3
New York	236	77	23	2	#	282	28	72	31	7
North Carolina	259	58	42	12	1	285	26	74	35	8
North Dakota	‡	‡	‡	‡	‡	292	14	86	42	7
Ohio	261	51	49	17	2	285	23	77	36	7
Oklahoma	255	59	41	6	2	275	33	67	22	3
Oregon	248	68	32	6	#	287	23	77	37	9
Pennsylvania	‡	‡	‡	‡	‡	287	22	78	39	8
Rhode Island	‡	‡	‡	‡	‡	277	33	67	28	5
South Carolina	‡	‡	‡	‡	‡	282	29	71	32	8
South Dakota	‡	‡	‡	‡	‡	289	18	82	39	7
Tennessee	‡	‡	‡	‡	‡	274	36	64	23	4
Texas	252	64	36	5	#	288	20	80	37	7
Utah	252	59	41	11	1	284	25	75	34	6
Vermont	‡	‡	‡	‡	‡	291	19	81	42	10
Virginia	263	48	52	15	4	288	22	78	38	9
Washington	243	71	29	5	1	287	23	77	38	10
West Virginia	‡	‡	‡	‡	‡	270	39	61	18	2
Wisconsin	260	53	47	12	3	287	23	77	38	8
Wyoming	‡	‡	‡	‡	‡	288	19	81	37	7
Other jurisdictions										
District of Columbia	226	85	15	2	#	249	65	35	8	1
DoDEA[1]	‡	‡	‡	‡	‡	286	21	79	34	5

‡ Reporting standards not met. Sample size is insufficient to permit a reliable estimate.

[1] Department of Defense Education Activity (overseas and domestic schools).

Note: The results for English language learners are based on students who were assessed and cannot be generalized to the total population of such students. Detail may not sum to totals because of rounding.

Source: U.S. Department of Education, Institute of Education Sciences, National Center for Education Statistics, National Assessment of Educational Progress (NAEP), 2007 Mathematics Assessment.

U.S. DEPARTMENT OF EDUCATION

The National Assessment of Educational Progress (NAEP) is a congressionally authorized project sponsored by the U.S. Department of Education. The National Center for Education Statistics, a department within the Institute of Education Sciences, administers NAEP. The Commissioner of Education Statistics is responsible by law for carrying out the NAEP project.

Margaret Spellings
Secretary
U.S. Department of Education

Grover J. Whitehurst
Director
Institute of Education Sciences

Mark Schneider
Commissioner
National Center for Education Statistics

THE NATIONAL ASSESSMENT GOVERNING BOARD

In 1988, Congress created the National Assessment Governing Board to set policy for the National Assessment of Educational Progress, commonly known as The Nation's Report Card™. The Governing Board is an independent, bipartisan group whose members include governors, state legislators, local and state school officials, educators, business representatives, and members of the general public.

Darvin M. Winick, Chair *President*
Winick & Associates
Austin, Texas

Amanda P. Aval lone, Vice Chair
Assistant Principal and Eighth-Grade Teacher
Summit Middle School, Boulder, Colorado

Francie Alexander
Chief Academic Officer, Scholastic, Inc.
Senior Vice President, Scholastic Education, New York, New York

David J. Alukonis
Chairman
Hudson School Board Hudson, New Hampshire

Barbara Byrd-Bennett
Executive Superintendent-inResidence
Cleveland State University, Cleveland, Ohio

Gregory Cizek
Professor of Educational Measurement
University of North Carolina, Chapel Hill, North Carolin

Shirley V. Dickson
Educational Consultant
Aliso Viejo, California

Honorable David P. Driscoll
Former Commissioner of Education
Massachusetts Department of Education
Malden, Massachusetts

John Q. Easton
Executive Director
Consortium on Chicago School Research
University of Chicago
Chicago, Illinois

Alan J. Friedman
Consultant
Museum Development and Science Communication
New York, New York

David W. Gordon
County Superintendent of Schools
Sacramento County Office of Education
Sacramento, California

Robin C. Hall
Principal
Beecher Hills Elementary School
Atlanta, Georgia

Kathi M. King
Twelfth-Grade Teacher
Messalonskee High School
Oakland, Maine

Honorable Keith King
Former Member
Colorado House of Representatives
Denver, Colorado

Kim Kozbial-Hess
Fourth-Grade Teacher
Hawkins Elementary School
Toledo, Ohio

In: Mathematics and Science Education
Editor: Chad P. Allerton

ISBN: 978-1-60692-313-9

Chapter 4

ANSWERING THE CHALLENGE OF A CHANGING WORLD: STRENGTHENING EDUCATION FOR THE 21ST CENTURY*

Margaret Spellings

INTRODUCTION: THE CHALLENGE—TO INNOVATE EDUCATION

We need to encourage children to take more math and science, and to make sure those courses are rigorous enough to compete with other nations. ... If we ensure that America's children succeed in life, they will ensure that America succeeds in the world
—President George W. Bush

High school reform is not just an "education issue." It's also an economic issue, a civic issue, a social issue and a national security issue. And it's everybody's issue.
—U.S. Secretary of Education
Margaret Spellings

America has long been innovation's home. It's in our very DNA, born from a desire to be free that was ahead of its time. When faced with a challenge, we invent the answer: from the first telephone to global satellite communications; from the first computer to the World Wide Web; from the Wright Brothers to Neil Armstrong. To Americans, innovation means much more than the latest gadget. It means creating a more productive, prosperous, mobile and healthy society. Innovation fuels our way of life and improves our quality of life. And its wellspring is education.

* U.S. Department of Education, Office of the Secretary, *Answering the Challenge of a Changing World: Strengthening Education for the 21st Century*, Washington, D.C., 2006.

President Bush has made innovation and education top priorities. He worked with Congress to pass the most far-reaching education reform in decades, the *No Child Left Behind Act* (*NCLB*). The law has brought high standards and accountability to public schools and sparked a mathematics and reading revival in the early grades. The president has also increased funding for innovative and intensive reading programs such as Reading First by more than 200 percent since 2001, benefiting more than a million students.

Our greatest advantage in the world has always been our educated, hard-working, ambitious people—and we're going to keep that edge.
—President George W. Bush

FACT:
In 2005, foreign-owned companies were a majority of the top 10 recipients of patents awarded by the U.S. Patent and Trademark Office.

The rest of the world, meanwhile, has not stood still. America no longer holds the sole patent on innovation. Inspired by our example, countries such as China, India and South Korea have invested heavily in education, technology and R & D. Billions of new competitors are challenging America's economic leadership. Our educational leadership has been challenged as well, with students from many developed nations outperforming ours in international tests, particularly in math and science. These test scores are linked to a lack of challenging course work, an ominous sign for many American schools. The impact may be felt well into the future: According to some estimates, America's share of the world's science and engineering doctorates is predicted to fall to 15 percent by 2010. [1]

This global challenge calls for bold action and leadership. America has done it before. Following the Soviet Union's 1957 launch of Sputnik, the world's first satellite, Congress passed and President Eisenhower signed into law the *National Defense Education Act of 1958*. The law accelerated the study of mathematics and science and helped improve foreign language teaching in our schools. It brought together the public and private sectors behind the effort. And it worked.

> *The Congress hereby finds and declares that the security of the Nation requires the fullest development of the mental resources and technical skills of its young men and women.*
>
> —National Defense Education Act of 1958

Within a decade, the number of science and engineering doctorates awarded in the U.S. annually had tripled, accounting for more than half the world's total by 1970.

Today, America faces not a streaking satellite but a rapidly changing global workforce. The spread of freedom is spurring technological innovation and global competition at a pace never before seen. We have to run to keep up. A high school diploma, once desirable, is now essential—and, increasingly, insufficient. About 90 percent of the fastest-growing jobs of the future will require some postsecondary education.[2] It is therefore unacceptable that among all ninth-graders, approximately three in 10 do not graduate on time, or that for black and Hispanic students the figure is about five in 10.[3]

Whether filling blue-collar or white-collar positions, employers now seek workers with "pocket protector" skills—practical problem-solvers fluent in today's technology. If current trends continue, by 2012 over 40 percent of factory jobs will require postsecondary education, according to the National Association of Manufacturers. And yet, almost half of our 17-year-olds do not have the basic understanding of math needed to qualify for a production associate's job at a modern auto plant.

> *U.S. manufacturing will no longer employ millions in low- skilled jobs. Tomorrow's jobs will go to those with education in science, engineering, and mathematics and to high-skill technical workers. Such a workforce is an important key to future growth, productivity, and competitiveness.*
>
> —National Association of Manufacturers, The Looming Workforce Crisis, 2005

Innovating and improving education is critical not only to America's financial security but also to our national security. Today, not one but 3,000 satellites circle the earth. U.S. soldiers use the latest technology and communications to fight the global war on terrorism. Advanced math skills are used to identify and undermine terrorist networks. Government and the private sector look to engineer new ways to protect lives and infrastructure from harm. And the effort to spread freedom to other nations and cultures demands speakers fluent in languages such as Arabic, Farsi, Chinese and Russian. Addressing these challenges will advance opportunity and entrepreneurship at home and promote democracy and understanding abroad.

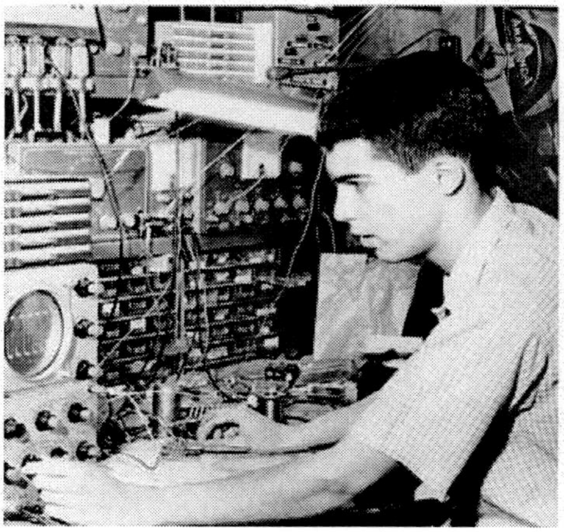

Rigorous instruction, high standards and accountability for results are helping to raise achievement in the early grades. Now America must complete the task. With our students working to achieve grade-level proficiency in math and reading by 2014, as called for by *NCLB*, innovative education reform is needed. America's civic, political and business leaders agree: To sustain our quality of life and way of life, we must act now. And President Bush is leading the charge.

FACT:

More than half of the undergraduate degrees awarded in China are in the fields of science, technology, engineering and math, compared to 16 percent in the U.S.

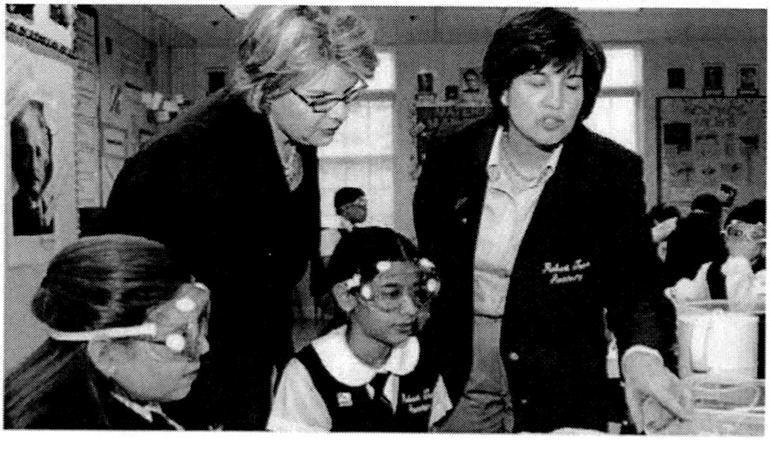

The Answer: President Bush's 2006 Education Agenda

President Bush's answer to America's newest challenge begins with the American Competitiveness Initiative. The American Competitiveness Initiative will commit $380 million in education funding and $5.9 billion overall in FY 2007 —and more than $136 billion over the next 10 years—to strengthen education, promote research and development and encourage entrepreneurship. The initiative will bring together leaders from the public sector, private sector and education community to better prepare our students for the 21st century. It will place a greater emphasis on math instruction from the earliest grade levels. It will ensure that high schools offer more rigorous course work, including Advanced Placement and International Baccalaureate programs in math, science and critical-need foreign languages. It will inform teachers of the most effective, research-based approaches to teaching math and science. It will encourage professionals in those fields to become teachers themselves. And it will coordinate federal math and science education programs to ensure the most effective use of the taxpayers' dollars.

The first bulwark in the face of rapidly changing economies and job markets is the flexibility and adaptability of the labor force. This adaptability begins with the formal educational system, especially the public schools.
—Ben Bernanke,
Federal Reserve Board chairman

The president's High School Reform Initiative ($1.475 billion in FY 2007) will help ensure that a diploma becomes a ticket to success for all graduates, whether they enter the workforce or go on to higher education. It will bring high standards and accountability to high schools by aligning their academic goals and performance with the No Child Left Behind Act. Through assessments and targeted interventions, the initiative will help educators raise achievement levels and close the achievement gap. It will also help alleviate the dropout problem by focusing more attention on at-risk students struggling to reach grade level in reading or math.

Finally, the president's National Security Language Initiative ($57 million in FY 2007), announced on Jan. 5, 2006, will help more American students master critical-need foreign languages to advance global competitiveness and national security. This joint project, in collaboration with the Department of State, Department of Defense and the Office of the Director of National Intelligence, will train teachers and assist students in those fields.

The inadequacies of our systems of research and education pose a greater threat to U.S. national security over the next quarter-century than any potential conventional war that we might imagine.
—U.S. Commission on National Security/21st Century (Hart-Rudman Commission), 2001

The Challenge: Knowledge of Math and Science

In this changing world, knowledge of math and science is paramount. "It's a magnificent time to know math," writes *Businessweek,* in an article entitled Math Will Rock Your World. So-called "math entrepreneurs" are translating the world into numbers—which translates into big salaries. According to the Bureau of Labor Statistics, job openings requiring science, engineering or technical training will increase by more than 24 percent, to 6.3 million, by 2014.

We must improve the way we teach math in our elementary schools. It's not just abouthelping younger students develop strong arithmetic skills; it's about planting the seeds of higher-order thinking for later in life.
—Secretary Margaret Spellings

Of all of the recommendations contained in the National Academies' report *Rising Above the Gathering Storm*, the highest priority is to vastly improve K–12 math and science education. Schools must help students develop the skills they will need to compete and succeed in higher education and the workforce, which are increasingly connected in this changed world. They must develop a pool of technically adept and numerically literate Americans to ensure a continual supply of highly trained mathematicians, scientists and engineers.

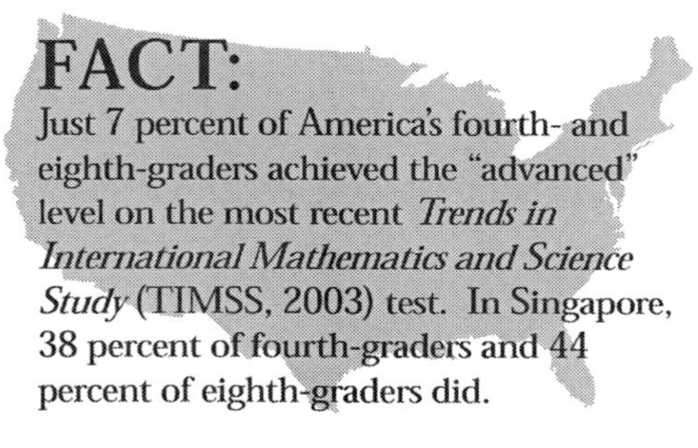

FACT:
Just 7 percent of America's fourth- and eighth-graders achieved the "advanced" level on the most recent *Trends in International Mathematics and Science Study* (TIMSS, 2003) test. In Singapore, 38 percent of fourth-graders and 44 percent of eighth-graders did.

We clearly have a long way to go. High school test scores in math have barely budged since the early 1970s. The percentage of seniors scoring below "basic" level on the National Assessment of Educational Progress assessments actually rose between 1996 and 2000 (see chart below). And less than half of high school graduates in 2005 were ready for college-level math and science course work, according to ACT.

How to read these

The percentages to the right of the shaded bars represent the percentages of students at or above Basic and Pro

The percentages in the shaded bars represent the percentages of sudents within each achievement level.

Math Performance of U.S. 12th-Graders

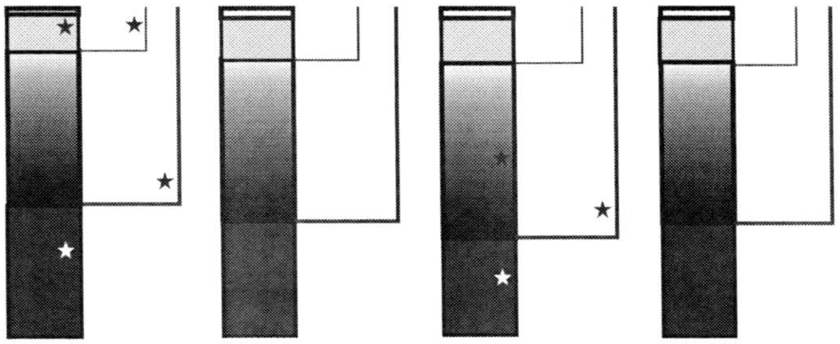

Signi cantly different from 2000.
NOTE: Percentages within each mathematics achievement level range may not add to 100, or to the exact percentages at or above achievement levels, due to rounding.
SOURCE: National Center for Education Statistics, National Assessment of Educational Progress (NAEP),1990,1992, 1996 and 2000 mathematics assessments.

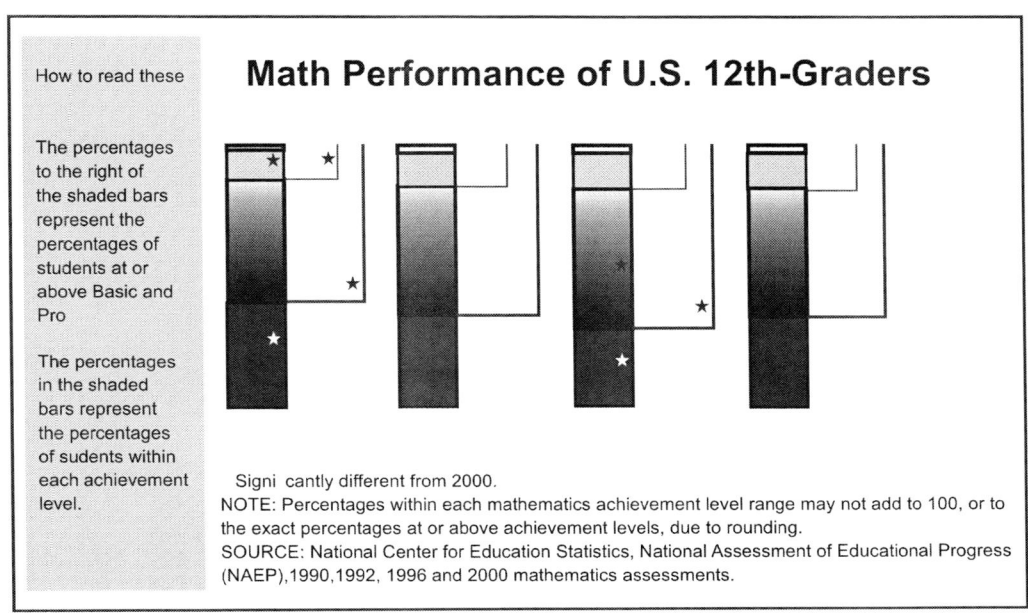

In 1983, the landmark *A Nation at Risk* report recommended that high school students be required to take a minimum of three years of math and three years of science to graduate. Yet today, only 22 states and the District of Columbia require at least this amount to graduate in the class of 2006. Even fewer require high school exit exams (which are often administered in 10th or 11th grade, leading many employers and universities to discount the results). Just one state—Alabama—calls for current students to take four years of both science and math to graduate.

The result is that as America's students grow older, the rest of the world catches up with them. Our 15-year-olds ranked 24th out of 29 developed nations in mathematics literacy and problem-solving, accord- ing to the 2003 Programme for International Student Assessment (PISA) test. The U.S. had a smaller percentage of top performers and a larger percentage of low performers than the average of all developed countries.

To succeed in [this] marketplace, U.S. firms need employees who are flexible, knowledgeable, and scientifically and mathematically literate.
—Norman Augustine, former chairman and CEO, Lockheed Martin; Chair, Committee on Prospering in the Global Economy of the 21st Century

A major part of the answer is teacher training. Three out of four fourth-grade math and science teachers in the U.S. do not have a specialization in those subjects. And students from low-income communities are far less likely to have teachers certified in the subject they teach. Shadowing it all, two-thirds of our math and science teachers are expected to retire by 2010, according to the National Commission on Mathematics and Science Teaching for the 21st Century. But talented teachers are always in great demand.

In this changing world, providing greater access to Advanced Placement courses is a must. AP courses benefit not only the students who take them but also the schools that offer them, whose curricula are upgraded and improved as a result. Students who take advanced math courses such as trigonometry, precalculus and calculus in high school are far more likely to earn a bachelor's degree in college. AP calculus students ranked first in the world on the Trends in International Mathematics and Science Study (TIMSS) test; U.S. students overall ranked second to last.

AP Course Participation, by Number of Courses

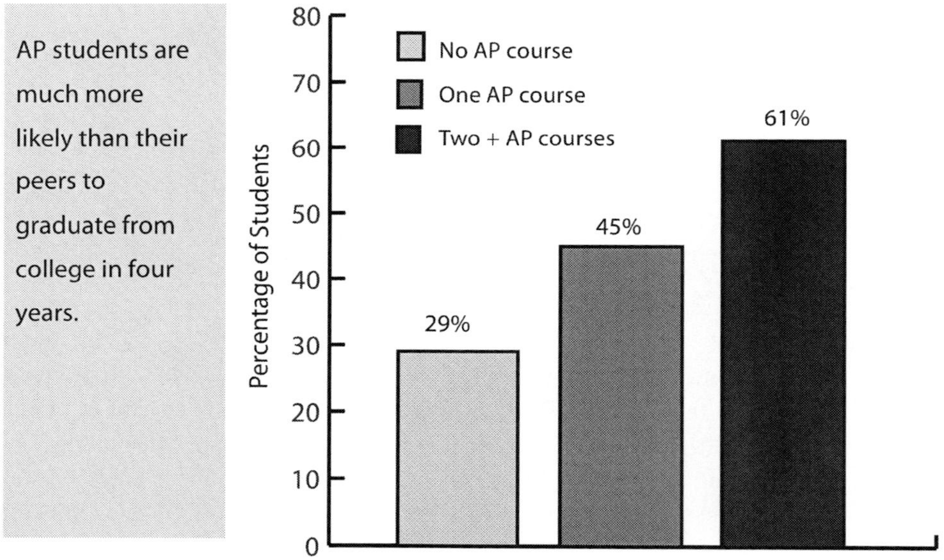

Source: College Board, 2006

Some progress has been made. Since 2000, the percentage of students who have taken and passed Advanced Placement courses has risen in all 50 states and the District of Columbia, according to the College Board. Even so, more than one-third of high schools—many serving predominantly low-income and minority students—still do not offer any AP courses. Based on PSAT scores, there were nearly a half million students in America ready for AP calculus last year who didn't take it or have access to it.

While many students eagerly await this opportunity, others still harbor negative views toward math and science. Old attitudes die hard. A recent survey commissioned by the Raytheon Company found that 84 percent of middle school students would rather clean their rooms, take out the garbage or go to the dentist than do their math homework.[4] According to the Business Roundtable, just 5 percent of parents say they would "try to persuade their child toward careers in science, technology, mathematics or engineering."[5]

In China today, Bill Gates is Britney Spears. In America today, Britney Spears is Britney Spears—and that is our problem.
—Thomas Friedman,
The World Is Flat, 2005

Many people still view math and science as "nerdy" subjects with little relevance to the "real world." Like it or not, that world has changed forever.

President Bush's American Competitiveness Initiative seeks to improve learning and instruction in mathematics and science through the following:

- **National Mathematics Panel** ($10 million): Modeled after the influential National Reading Panel, the National Math Panel would convene experts to empirically evaluate the effectiveness of various approaches to teaching math, creating a research base to improve instructional methods for teachers. It would lay the groundwork for the Math Now program for grades K–7 to prepare every student to take and pass algebra;
- **Math Now for Elementary School Students** ($125 million): Like the successful and popular Reading First program, Math Now for Elementary School Students would promote promising, research-based practices in mathematics instruction and prepare students for more rigorous math course work in middle and high school;
- **Math Now for Middle School Students** ($125 million): Similar to the current Striving Readers Initiative, Math Now for Middle School Students would diagnose students' deficiencies in math proficiency and provide intensive and systematic instruction to enable them to take and pass algebra;

- **Advanced Placement-International Baccalaureate (APIB) Incentive Program** ($122 million—$90 million over 2006 levels): The AP-IB Incentive Program would train 70,000 additional teachers to lead AP-IB math, science and critical-need language courses over the next
- five years. It would increase the number of students taking AP-IB tests to 1.5 million by 2012, tripling the number of passing test-takers to approximately 700,000 while giving them the opportunity to earn college credit;
- **Adjunct Teacher Corps** ($25 million): The Adjunct Teacher Corps would encourage by 2015, 30,000 qualified math and science professionals to share their knowledge as adjunct high school teachers.
- **Evaluating the Effectiveness of Federal Science, Technology, Engineering and Math (STEM) programs** ($5 million): A governmentwide effort would be undertaken to determine which federal education programs are most effective in raising achievement in math and science, which deserve more funding and which should be consolidated to save taxpayer money. The initiative would also align these education programs with the goals and aims of the *No Child Left Behind Act.* Thirteen agencies reported spending $2.8 billion on 207 education programs in FY 2004. About half of the programs dedicated to math and science received less than $1 million in funding, with most targeted to postsecondary education; and
- **Including Science Assessments in *NCLB***: *NCLB* requires every state to develop and administer science assessments once in each of three grade spans by the 2007–08 school year. Including these assessments in the accountability system will ensure students are learning the necessary content and skills to be successful in the 21st-century workforce.

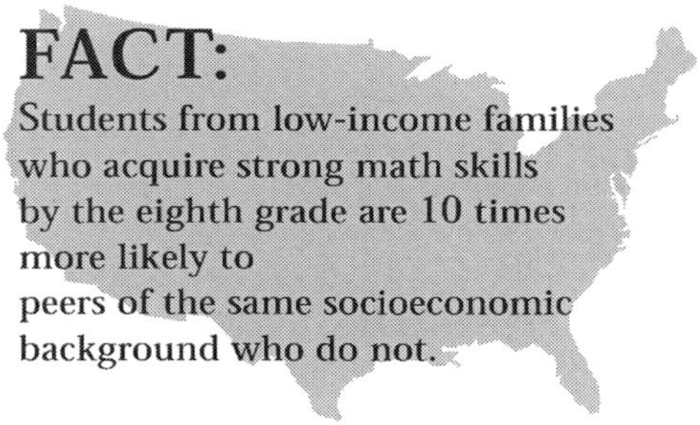
FACT:
Students from low-income families
who acquire strong math skills
by the eighth grade are 10 times
more likely to
peers of the same socioeconomic
background who do not.

One of the best standard predictors of academic success at Harvard is performance on Advanced Placement examinations.
—Bill Fitzsimmons, Dean of Admissions and Financial Aid, Harvard University

Other Math and Science Initiatives

- **Academic Competitiveness grants** and **SMART Grant Program** ($850 million in FY 2007): Approved by Congress, these two higher education grant programs build on the success of the Pell Grant Program and will benefit more than 500,000 students in need.

Academic Competitiveness grants will provide increased funds for low- income students who take a rigorous academic curriculum in high school. Grants in the amount of $750 would be awarded to qualified first-year college students who completed a rigorous high school program; grants in the amount of $1,300 would be awarded to second-year students who completed a rigorous program and who earn and maintain a 3.0 average in college.

SMART grants, in the amount of $4,000, will go to college juniors and seniors studying math, science or critical-need foreign languages who also earn and maintain a 3.0 GPA. This will encourage more students to go into fields that improve America's security and competitiveness.

- **Mathematics and Science Partnerships** ($182 million in FY 2007): This program supports the American Competitiveness Initiative by providing state formula grants to help improve students' academic achievement in rigorous math and science courses. It also assists teachers by integrating proven, research-based teaching methods into the curricula.
- **Expanded Teacher Loan Forgiveness**: This popular program offers up to $17,500 (up from $5,000) in loan forgiveness for highly qualified math, science and special education teachers serving challenging, low- income schools and communities.

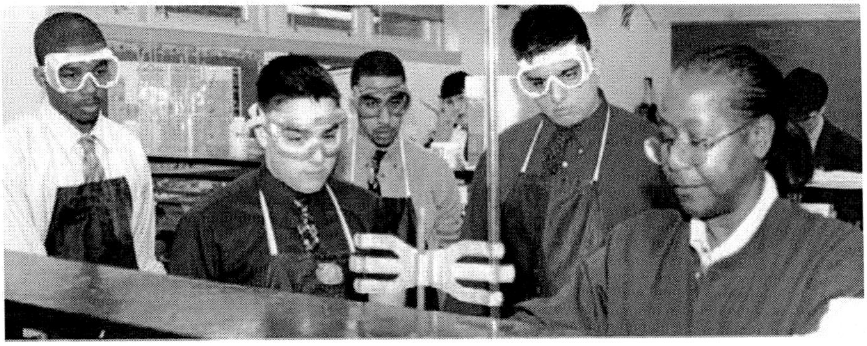

THE CHALLENGE: ACCELERATING OUR SCHOOLS' PROGRESS

Innovating and improving America's schools will not occur overnight. It took time for eight other developed nations to surpass America's high school graduation rate among adults aged 25 to 34; and it will take time for the U.S. to regain its leadership. We must start by accelerating our progress.

A national problem demands a comprehensive solution. As Secretary Spellings noted, "The competition starts in elementary school." The good news is that educators and policymakers are learning more and more about what works. A half-century ago, the U.S. turned the threat of Soviet competition into proof of our ability to improve our schools and quality of life. Just four years ago, the U.S. turned a growing achievement gap into the bipartisan No Child Left Behind Act.

Thanks to the [No Child Left Behind] law, which requires states to assess children annually and to break down the results by minority group and income level, it has for the first time become possible to track which schools are failing which students. More important, the law also requires states to turn schools around and help them succeed.

—Editorial, *Washington Post*, Feb. 5, 2006

The law set a course for all students to attain grade-level proficiency in the core subjects of reading and math by 2014. Students in grades 3 through 8 are now learning under high standards. Teachers are using proven instructional methods. Schools are being held accountable for results. Parents have more information and choices. And states have more flexibility to spend federal K–12 education funds, which have increased by 40 percent since 2001.

The early results are in. Across the country, academic achievement has risen significantly in the earliest grades, with math scores at all-time highs, including among African-American and Hispanic students. In the last two years, the number of fourth-graders who learned their fundamental math skills increased by 235,000—enough to fill 500 elementary schools. More reading progress was made among 9-year-olds over the last five years than in the previous 28 years combined. Meanwhile, according to the Nation's Report Card, the achievement gaps in reading and math between white and African-American 9-year-olds and between white and Hispanic 9-year-olds are at all-time lows. Educators use terms like "amazing," "stunning" and "remarkable" to describe the progress.

We've seen the results in my own state of Massachusetts. Student achievement is up across the board in both reading and math. We've made significant progress in educating children with disabilities. We're seeing the results of the No Child Left Behind reforms in other parts of the country, too. ... Research-based instruction, assessments, and targeted interventions are working.

—Sen. Edward Kennedy (D-Mass.)

No Child Left Behind has set the goal of every child achieving, but the states and schools themselves have done the heavy lifting to make the law work. For the first time, all 50 states have unique accountability plans in place, with real consequences attached. The results can be seen in schools like Maryland's North Glen Elementary. In 2003, just 57 percent of North Glen's students were proficient in reading, while 46 percent were proficient in math. Those numbers have skyrocketed to 82 percent and 84 percent, respectively. As First Lady Laura Bush said during a recent visit to North Glen, "They've taken advantage of all the aspects of the *No Child Left Behind* law."

Another example is Charles L. Gideons Elementary School in Atlanta. The number of its students meeting Georgia's standards in reading increased by 23 percentage points since 2003. For math the news is even better: a 34 percentage-point improvement during the same period.

A districtwide success occurred in Garden Grove, Calif. Three-fourths of the Garden Grove Unified School District's students did not speak English. Nearly 60 percent were from low-income families. Nevertheless, all but two of the district's 67 schools met or exceeded their Adequate Yearly Progress goals under the law.

We use No Child Left Behind to set the targets we want to hit. We align all our actions and resources to hit those targets. And we believe the kids can do it.

—Laura Schwalm, superintendent, Garden Grove Unified School District

The *No Child Left Behind Act* was designed to improve achievement. But it has also shown us what is achievable. Educators, administrators and public officials are working together, united behind a worthy goal. Now it's time to apply the act's successful principles to our nation's high schools.

Every kid can graduate ready for college. Every kid should have the chance. Let's redesign our schools to make it happen.

—Bill Gates, chairman, Microsoft Corporation

There is not a moment to waste. Governors and business leaders are united in calling for urgent reform. Every year approximately one million students drop out of high school, costing the nation more than $260 billion dollars in lost wages, taxes and productivity over the students' lifetimes. A high school graduate can expect to earn about $275,000 more over the course of his or her lifetime than a student who doesn't finish high school; a college graduate with a bachelor's degree can expect to earn about $1 million more. Dropouts are also three-and-a-half times more likely to be arrested, according to reports. Encouraging at-risk students to stay in school by addressing their academic needs will improve their quality of life and that of their fellow Americans.

THE ANSWER: THE HIGH SCHOOL REFORM INITIATIVE $1.475 BILLION IN FY 2007

Giving Americans the math and science skills they need will help us remain a world economic leader. Teaching students under the highest standards and offering more rigorous coursework will help us remain an education leader.

The president's **High School Reform Initiative** would hold high schools accountable for providing high-quality education to all students. And it would help educators implement strategies to meet the needs of at-risk high school students. The proposed program would make formula grants to states to support:

- The development, implementation and evaluation of targeted interventions designed to improve the academic performance of students most at risk of failing to meet state academic standards; and
- Expanded high school assessments that would assist educators in increasing accountability and meeting the needs of at-risk students.
- Interventions would be designed to increase the achievement of high school students; eliminate achievement gaps between students from different ethnic and racial groups and income levels; and help ensure that students graduate with the education, skills and

knowledge necessary to succeed in postsecondary education and in the technology- based global economy.

FACT:

Only one in five recent high school graduates in the workforce say they were challenged with high academic expectations in high school, according to Achieve, Inc.

A key strategy would be the development of individual performance plans for students entering high school, using eighth-grade assessment data in consultation with parents, teachers and counselors. Specific interventions may include programs that combine rigorous academic course work with vocational and technical training, research-based dropout prevention activities, and the use of technology-based assessment systems to closely monitor student progress. In addition, programs that identify at-risk middle school students for assistance would help prepare them to succeed in high school and enter postsecondary education. This includes college preparation and awareness activities for students from low- income families.

The president's proposal also would require states to develop and implement reading and mathematics assessments in two additional grade levels in high school, building on the current *NCLB* requirement for testing once in grades 10–12. The new assessments would inform strategies to strengthen school accountability and meet the needs of at-risk students.

Additional Support

Striving Readers ($100 million in FY 2007): First funded in 2005, this program would be expanded significantly to reach more secondary students reading below grade level, which puts them at risk of dropping out. Students would benefit from research-based interventions coupled with rigorous evaluations. Schools also would benefit from activities and programs designed to improve the overall quality of literacy instruction across the entire curriculum.

THE CHALLENGE:
PROMOTING FREEDOM AND UNDERSTANDING

America faces a severe shortage of people who speak languages that are crucial to its national security and global competitiveness:

- According to the Center for Applied Linguistics, less than one- fourth of public elementary schools report teaching foreign languages, even though a child's early years are the best years in which to learn a new language.
- Less than 1 percent of American high school students study Arabic, Chinese, Farsi, Japanese, Korean, Russian or Urdu—combined.
- Less than 8 percent of undergraduates in American universities take foreign language courses, and less than 2 percent study abroad in any given year.

To prepare young Americans to understand the peoples who will help to define the 21st century, nothing is more important than our ability to converse in their native tongues.
—U.S. Secretary of State Condoleezza Rice

While only 44 percent of U.S. high school students were studying a foreign language in 2002, learning a second or even a third foreign language is compulsory for students in the European Union, China, Thailand and elsewhere.

More than 200 million children in China study English. By comparison, only about 24,000 elementary and secondary school children in the U.S. study Chinese. Many students in other nations begin learning another language before they are even 10 years old. They will have an edge over monolingual Americans and others in developing new relationships and business connections in countries other than their own.

The Answer: National Security Language Initiative $57 Million in FY 2007

Critical-need foreign language skills are necessary to advance the twin goals of national security and global competitiveness. Together with the Department of State, Department of Defense and the Office of the Director of National Intelligence, the Department of Education will propose to offer grants and training for teachers under President Bush's National Security Language Initiative.

The initiative will increase the number of Americans who speak and teach foreign languages, with an emphasis on critical-need languages. It will strengthen and refocus the federal Foreign Language Assistance Program, and will initially enable 24 school districts across the country to create partnerships with colleges and universities to develop critical-need language programs. Among the critical-need languages identified are Arabic, Chinese, Korean, Japanese, and Russian, as well as languages in the Indic, Iranian and Turkic families.

The National Security Language Initiative will also provide funding to create a Language Teacher Corps, with the goal of having 1,000 new foreign language teachers in U.S. schools by the end of the decade. And it will enable the creation of the e-Learning Language Clearinghouse and expanded Teacher-to-Teacher seminars to assist foreign language teachers anytime, anywhere.

Conclusion

Thanks to our schools, the 20th century was known as the "American Century." The 21st century remains to be claimed. But Americans have never backed down from a challenge. This changing world offers another opportunity for Americans to shine, and the President's American Competitiveness Initiative and the rest of his 2006 education agenda will help set the course.

In the 21st century, economic power will be derived from skills and innovation. Nations that don't invest in skills will weaken; it is that straightforward.
—Louis Gerstner Jr., former chairman and CEO, IBM

America's schools and students have made great progress in improving academic achievement in the early grades. But like athletes or musicians, children of all ages must work hard each and every day if they wish to compete, perform and succeed—and their schools must show them the way. The president's 2006 education agenda will help prepare the students of today to become the successful leaders—the pioneers, discoverers and Nobel Prize winners—of the next American Century.

PRESIDENT'S AMERICAN COMPETITIVENESS INITIATIVE: EDUCATION *FY 2007 Budget Request* Math Panel $ 10 million	
Math Now for Elementary Students	$125 million
Math Now for Middle School Students	$125 million
AP-IB	$122 million *
Adjunct Teacher Corps	$ 25 million
Evaluation of Federal STEM programs	$5 million

* includes $32 million for AP-IB from FY 2006

REFERENCES

[1] The estimate of America's share of the world's science and engineering doctorates is from "Does Globalization of the Scientific/Engineering Workforce Threaten U.S. Economic Leadership?", by Richard B. Freeman, in *Working Paper Series,* National Bureau of Economic Research: Working Paper 11457, June 2005, page 5.

[2] The estimate of the percentage of the fastest-growing jobs that will require postsecondary education is from remarks by U.S. Secretary of Labor Elaine L. Chao, on Feb. 22, 2006, at the WIRED Initiative Town Hall Event in Washington, D.C.

[3] The estimates of graduation rates for all, black and Hispanic ninth-graders are from "Public High School Graduation and College-Readiness Rates: 1991-2002," by Jay P. Greene and Marcus A. Winters, Manhattan Institute for Policy Research, New York: Education Working Paper No. 8, February 2005.

[4] The survey by the Raytheon Company regarding middle- schoolers' attitudes toward math is from www.raytheon.com/ about/contributions/mathmovesu (accessed April 4, 2006).

[5] The Business Roundtable study, *Innovation and U.S. Competitiveness: Addressing the Talent Gap*, 2005, is at www. businessroundtable.org/pdf/20060112Two-pager.pdf (accessed April 4, 2006).

INDEX

A

B

C

D

E

I

J

K

L

M

N

O

P

Q

R

S

Y

Z